T0236971

Lecture Notes in Computer Science 11690

More information about this series at http://www.springer.com/series/7407

Bob Coecke · Ariane Lambert-Mogiliansky (Eds.)

Quantum Interaction

11th International Conference, QI 2018
Nice, France, September 3–5, 2018
Revised Selected Papers

 Springer

Editors
Bob Coecke 🆔
University of Oxford
Oxford, UK

Ariane Lambert-Mogiliansky 🆔
Paris School of Economics
Paris, France

ISSN 0302-9743   ISSN 1611-3349   (electronic)
Lecture Notes in Computer Science
ISBN 978-3-030-35894-5   ISBN 978-3-030-35895-2   (eBook)
https://doi.org/10.1007/978-3-030-35895-2

LNCS Sublibrary: SL1 – Theoretical Computer Science and General Issues

This Springer imprint is published by the registered company Springer Nature Switzerland AG
The registered company address is: Gewerbestrasse 11, 6330 Cham, Switzerland

Preface

Quantum Interaction (QI) is an emerging interdisciplinary field of science. It proposes applications of quantum theory to a large variety of domains from psychology, economics, semantic and memory, natural language processing, cognition, information retrieval, biology, and political science. The applications addressed typically operate at a macroscopic scale and could not be considered quantum in a quantum mechanical sense. However, they share key properties with quantum systems. These include non-commutativity of measurement, indeterminacy, non-separability, contextuality, and harmonic oscillations. QI thus refers to the use of the quantum mathematical, conceptual, or probabilistic structures outside of physics. Since its inception in 2007, QI has evolved from nearly exclusively theoretical and conceptual contributions to more applied works including lab experiments.

QI 2018, the 11th International Conference on Quantum Interactions was part of a series of international conferences. This now traditional conference started in 2007 as part of the Association for the Advancement of Artificial Intelligence (AAAI) Spring Symposia at Stanford University. For the second time the conference was held in France (QI 2012 was in Paris). The 11th conference took place during September 3–5, 2018, in Nice. It was hosted by Nice Graduate School of Management at the Sophia Antipolis University.

In this year's conference we had many distinguished speakers, and we are happy to have contributions to this volume from two of our keynote speakers. Prof. Michel Bitbol, Directeur de recherche CNRS at Ecole Normale Supérieure, Paris, France, offers a philosophical and historical perspective that unveils deep reasons for a quantum approach in human sciences. Prof. B. P. F. Jacobs, from the Institute for Computing and Information Sciences (ICIS), Correctness Digital Security Group at Radboud University Nijmegen, The Netherlands, provides a review of the questions related to updating in the classical and quantum context and introduces challenging research issues related to probabilistic logic.

The conference had an affiliated workshop entitled Workshop on Compositional Approaches in Physics, NLP, and Social Sciences (CAPNS 2018). Its proceedings appeared in ENTCS.

We would like to take the opportunity to thank everybody who made this symposium possible: the Steering Committee, the Program Committee members for their reviewing job, the proceedings and the publicity chairs, those responsible for the website design and management, and all the conference participants and presenters. We are grateful for the support given by the University of Nice Sophia Antipolis.

June 2019

Bob Coecke
Ariane Lambert-Mogiliansky

Organization

Program Chairs

Bob Coecke   University of Oxford, UK
Ariane   Paris School of Economics, France
    Lambert-Mogiliansky

Steering Committee

Peter Bruza   Queensland University of Technology, Australia
Trevor Cohen   University of Texas at Houston, USA
Bob Coecke   University of Oxford, UK
Ariane   Paris School of Economics, France
    Lambert-Mogiliansky
Dominic Widdows   Grab Technologies Inc., USA

Program Committee

Irina Basieva   LNU, Sweden
Bob Coecke   University of Oxford, UK
Trevor Cohen   University of Texas, Houston, USA
Ehtibar Dzhafarov   Purdue University, USA
Bart Jacobs   Radboud University, The Netherlands
Chris Heunen   The University of Edinburgh, UK
Dominic Horsman   University of Durham, UK
Dimitri Kartsaklis   Apple Inc., UK
Kirsty Kitto   University of Technology, Sydney
Andrei Khrennikov   Linnaeus University, Sweden
Ariane   Paris School of Economics, France
    Lambert-Mogiliansky
Bill Lawless   Paine College, USA
Martha Lewis   University of Amsterdam, The Netherlands
Dan Marsden   University of Oxford, UK
Simon Perdrix   CNRS, University of Grenoble, France
Emmanuel Pothos   City University London, UK
Quanlong Wang   University of Oxford, UK

Local Committee

Sébastien Duchêne   University of Nice Sophia Antipolis, France
Ismaël Rafaï   University of Nice Sophia Antipolis, France
Laurence Gervasoni   University of Nice Sophia Antipolis, France

Contents

Fundamentals

Why Should We Use Quantum Theory?
The Case of Human Sciences

Michel Bitbol[(✉)]

CNRS, Archives Husserl, Ecole Normale Supérieure,
45, rue d'Ulm, 75005 Paris, France
michel.bitbol@ens.fr

1 Introduction

Why is quantum theory so universal? Why does it apply to so many situations beyond the field of microphysics? To answer such questions, we can find inspiration from a remarkable reflection of the French philosopher of Science Jean Cavaillès about probabilities: "If any physical law is nothing else than a gamble for action, the scandal of probabilities ceases: far from being an inadequate substitute for our power to know, probabilities must be seen as the paradigm and the foundation of all scientific activity" [6]. Accordingly, if quantum laws are understood as a gamble for action, the "scandal" of their probabilistic status ceases. The careful analysis of the structure of probabilistic valuations based on state vectors in a Hilbert space is capable of profoundly transforming our view of quantum mechanics. It can help us to suspend the relationship of distrust (not to say of distance and strangeness) that we have maintained with this theory since its creation, and to change it into a feeling of proximity and trust. It can help us see quantum mechanics, not as an anomaly in the space of physical theories, but rather as their deepest archetype.

From this decidedly probabilistic standpoint, the application of quantum theory to many cases in the human sciences is no longer a coincidence, but a necessity. That it is a necessity becomes clear as soon as one understands two things: (i) that quantum theory formalizes a broad class of activities of knowledge rather than the objects of such knowledge; and (ii) that there is an isomorphism between certain situations of knowledge typical of the human sciences, and most situations of knowledge in microphysics. To put things shortly, both the human sciences and microphysics deal with situations in which knowing is not tantamount to observing, but rather to intervening and participating. As soon as we realize that quantum theory is above all a generalized probabilistic valuation for situations of knowledge in which intervention and participation are insurmountable, its universality and its applicability to such cases as decision-making, case-sensitive categorization and behavior in situations of uncertainty, becomes almost trivial.

By contrast, the purely pragmatic justification that is usually provided for applying quantum theories to the human sciences looks shy and contrived. The standard name for this application is "quantum model": quantum model of decision, quantum model of perception, quantum model of meaning ascription etc.

© Springer Nature Switzerland AG 2019
B. Coecke and A. Lambert-Mogiliansky (Eds.): QI 2018, LNCS 11690, pp. 3–21, 2019.
https://doi.org/10.1007/978-3-030-35895-2_1

And the best reason one then finds for the ("surprising") validity of such models is that quantum theories are "rich" and that "A richer expression scheme comes with a greater modeling power" [27].

This looks weak yet reasonable. But is the status of "model" ascribed to quantum theoretical accounts of human processes credible ? First of all, what about this word "model", that mostly refers to restrictive uses of a theory aiming at making sense of some particular experimental situations ? Remind that "model" is a word derived from the latin 'modulus', 'a small measure'. Its primary semantic content is therefore a *likeness made to scale*. But what if, as Heisenberg and several other creators of quantum mechanics suspected, quantum theory is no image, and no likeness, of anything ? What if quantum theory does not resemble its object in any way, but only formalizes gambles about the reactions of any object whatsoever when it is put in a certain class of situations of knowledge [11] ? What if, as Richard Healey pointed out, "quantum theory makes a radical break with previous physics not because of the weirdness of the physical behavior it represents, but ... (because) quantum theory is simply not in the business of representing what happens in the physical world" [15] ? Then, the very idea of quantum models representing some human processes is undermined. But conversely, the fact that quantum theory can indeed be applied to human sciences makes a non-representationalist interpretation of quantum mechanics much more compelling.

2   On the Human Science/Natural Science Issue

There are some philosophical reasons to this persistent resistance to quantum theoretical approaches of the human sciences. Those who resist sometimes fear that the desire to formulate quantum theories of human processes stems from an old thesis proposed by Otto Neurath around 1930 under the name of "Physicalism". The latter thesis amounts to a quest for the general unity of science under the exclusive authority of physics promoted "queen of sciences". But the search for the unity of the sciences by their absorption in physics understood as universal knowledge may well imply a reductionist act of faith, with the basic assumption that only the things described by physics are real, and that all the rest, especially the individual and social processes that concern the human sciences, is only an epiphenomenon. John Searle expressed this conflation of physicalism and reductionism thus: "That," he said, "is the raw structure of our ontology. We live in a world made entirely of physical particles in fields of forces. Some of them are organized into systems (...). Now the question is: how can social facts be justified within this ontology?" [26]. Here, physics reveals ontology; ontology holds the ultimate truth about the world; and every other science, including social and mental sciences, must conform to it.

This way of relating quantum mechanics to the human sciences certainly represents a possibility for thought, but it is not the only one, and not even the most interesting. Another philosophical approach may lead us to identify a deeper relationship between the two families of sciences. This is the approach, familiar to Bas Van Fraassen, which consists in giving the *methods* of scientific research an importance and scope greater than that of the contents of scientific representations. Scientific representations are historically shifting, ontologies replace one another, but the methods of science continuously become broader and more refined. It may be useful, as an exercise in self-consistency of the system of science, to articulate the representations associated to physics with those associated to the human sciences. But it is much more instructive to identify the common constraints that eventually lead to methodological prescriptions shared by advanced physical theories and the human sciences. On the basis of these shared methodological prescriptions, we can indeed arrive at unified forms of theorization, even though there is no reason to mix up the objects of the two sciences, and no reason either to believe in the possibility of reducing them to each other.

Let us then examine the methods of modern physics and the human sciences; Let's try to identify what they have in common, and above all what are the constraints that shaped them in so similar ways.

It should be noted, even before beginning the investigation, that the methodological similarity with the human sciences does not concern classical physics. As long as one sticks to classical physics and its epistemological paradigm, one can accept the strict dichotomy between natural sciences and human sciences that the hermeneutic tradition has constantly affirmed since Dilthey. According to this dichotomy, the natural sciences yield *explanations* of the processes they study, whereas the human sciences are based on the possibility for a human subject to *understand* her co-subjects, that is to say to *simulate* their inner states, to imagine what she would have done if she had been "in their shoes". The sciences of nature *explain* an environment which we are free to contemplate at a distance, while the human sciences explore our possible insertions in the situation and in the (instrumental and linguistic) practices of our fellow human beings. Habermas reformulated this difference as follows: "The theories of the natural sciences are systems of statements about states of affairs, whereas, in the human sciences, the complex relation between statements and states of affairs is already present in the states of affairs under analysis" [14]. According to this version of the traditional distinction, the natural sciences manage to completely separate their object from their means of study and linguistic designation. On the contrary, the human sciences are condemned to maintain an inextricable link between what they aim to study (the "states of affairs") and their interpretative grid conveyed by language (the "statements"). Indeed, the human sciences have what some consider to be a disadvantage compared to the classical sciences of nature: their act of investigation is coextensive with the field investigated; their own personal, social, and linguistic processes partake of what they seek to know.

There is also a second distinction, stated by Von Kempski and endorsed by Habermas: "The studies of the social sciences (...) are essentially studies of *possible* action, whereas theoretical physics always refers to actual nature" [14]. In other words, the classical sciences of nature describe what happens, what is the case, whereas the human sciences content themselves with identifying what holds potentialities for events to come. The reason is that the human sciences are not so much concerned with the present *causes* as with the *reasons* for acting in the future; not so much on facts as on norms. As Jean Piaget writes, "The three fundamental notions to which social structures are reduced are those of rules, of values and of signs; and at first sight they seem irreducible to the concepts used in the natural sciences" [23]. A widespread tendency to mix up the two approaches, could even be considered as the root of many false riddles about the mind and body: it is this kind of confusion that Gilbert Ryle called a "category mistake" [25].

However, the two major epistemological differences that have just been identified between the science of nature and the human sciences are partially canceled when the natural sciences adopt the paradigm of quantum physics instead of the paradigm of classical physics.

Firstly, we know that quantum physics is working on an experimental material in which the separation between what is explored and the means of exploration, meets fundamental limits. Like the human sciences, quantum physics must take into account the involvement of practices of knowledge in the phenomenon to be known. Like the human sciences, quantum physics deals with a situation where the epistemic act is coextensive with its field of study.

Secondly, because of the limits to the separation between the domain explored and the means of exploration, quantum physics can no longer claim to describe the properties of entities, as if it were from outside. All it can specify is the evolution of a pre-probabilistic symbol (the state vector) that can be used as an internal guideline for experimental and technological research. Just as the human sciences, quantum physics concentrates its theoretical activity on identifying potentialities for guiding effective operations, and no longer on highlighting actual events that are supposed to be independent of any intervention. The human sciences and quantum physics thus have in common a crucial constraint which they must take into account, and which is likely to be reflected in their methods as well as in the structure of their theoretical products.

3   The Human Roots of Quantum Science

The analogy with the human sciences was first noticed by Bohr, and it seems to have guided his research in physics. Testimonies show that his early familiarity with the psychology and philosophy of knowledge of the Danish thinker Harald Höffding determined his interpretation of quantum theory [22]. Instead of quantum theory playing the role of ontological "foundation" for the sciences, including human sciences, it is the human sciences that, through the voice of Höffding, offered to the nascent quantum theory a model of non-standard epistemological positioning.

Let's now examine this alternative model. Höffding's psychology had among its most fundamental principles a "law of relations" stipulating that no mental state can be considered as existing in itself, regardless of the relations it has with other mental states. Very soon, this psychological "law of relations" was extended by Höffding to physical properties, giving Bohr an example of how to export thought patterns from one science to another. According to Höffding, in the same way as mental states are relational rather than monadic traits, "(the) 'things' (...) are only understood as their properties, and the properties are manifested as so many relations to other stuff" [19]. Molecules, atoms and electrons are still 'things', totalities; but they are only known and understood as relations. A particular interest of the "law of relations", according to Höffding, was that it manifested, with particular force in the mental domain, a universal fact of interdependence between the knowing and the known. In the mental domain, more than in any other, the relation of knowledge comes first, and the related terms (subject and object) are derived. Rather than admitting without discussion an object and a subject given in advance, one realizes that each act of knowledge relates "(...) an objectively determined subject with a subjectively determined object" [18]. In other terms, subject and object are correlative of each other in each act of knowledge; they help to define each other through a process of delimitation of one by the other. In a style that evokes Hegel's dialectic, Höffding therefore asserts that the progress of knowledge is conditioned by a mutual revelation of the subjective part in the object, and of the objectivable part of the subject.

On careful examination, it is quite easy to see this dynamic at work in history. The most important advances of the sciences of nature have had as a prerequisite a revelation of the function fulfilled by the situation of the knowing subject in the phenomena that were previously supposed to be "objective" and therefore independent of knowledge. This was the case when Copernicus connected the apparent motion of the planets to their relation to an astronomical subject living on the planet Earth, or when Galileo connected the speed of material bodies to their relationship with an observer who is conventionally motionless. But this condition of subjectivation of the objective field is accompanied by its converse, which is the objectification of the subjective. Thus, all that Copernicus has retained from the subject is a terrestrial location, and all that Galileo has retained from the subject is an origin of the geometrical reference frame.

It should be noted at this stage that the co-definition of the objective and subjective sides of knowledge, as a consequence of their origin in an inextricable epistemic relation, has as its correlate the incompleteness of their characterization, and their being in constant development. For each revision of the polarity of the subject-object link, a redefined object arises before a subject whose questioning is renewed accordingly.

Here again, the human sciences and especially psychology serve as an emblematic case, because they are confronted with the most extreme version of the knower's inextricability in the process of knowledge. Höffding notices that, because it tries to objectify its field of study, the introspective psychology of his time misses a crucial aspect of mental activity: the possibility of being so immersed in it, that no reflection on it is allowed. It is impossible to have desire, and to analyze one's own motivations at the same time. It is impossible to be attentive, and to be attentive to this attention at the same time. The object of attention is constantly redefined as the attentive subject reorients herself. Another way to express this is the following: when the subject of attention is not really distinct from its object, the act of observation can only alter the latter. This is true in introspective psychology, since when I attempt to observe the subject that I am, I transform myself into another subject capable of positing this very attempt as a new object of my attention. And this is also true in experimental psychology, since the subject being studied is likely to reshape itself according to the experimenter's expectations. What should we do, faced to this dilemma? What should we choose, between a continuous absorption into one's own subjective experience, and the systematic objectifying distance advocated by scientific psychology? In the wake of his understanding of subject and object through a dynamic of reciprocal definition, Höffding advises not to priviledge one of these two mutually exclusive attitudes; he advises not to remain stuck in either commitment or distancing. Both attitudes are indispensable in some ways. Their incompatibility does not exclude their alternative use.

Here, it is not difficult to recognize the conceptual structure that Bohr named "complementarity". Complementarity in Bohr's sense refers to the joint use of two mutually exclusive characterizations of the same object.

In quantum physics, Bohr identified a reason for the relevance of the concept of complementarity: it is the finite character of the quantum of action, which comes into play at the moment of the measurement process. The indivisibility of the elementary quantum exchanged at the moment of the interaction between a microscopic system and a measuring device does not allow to separate, in the phenomenon, the contribution of the system from the contribution of the apparatus. But if phenomena cannot be decomposed into a contribution of the object and a contribution of the apparatus, no inference from them to an "independent" object can be done. The microscopic phenomenon is nothing more than the expression of a global experimental situation; it is not the revelation of an intrinsic property of some object.

Bohr also notes that it remains necessary to describe part of the measuring instruments by means of current language appropriately refined by the terminology of classical physics. Indeed, this immediately significant part of measuring instruments must be interpreted in terms of objects with definite "properties", if its indications are to be communicated in an unambiguous way to all experimenters by means of propositions containing a grammatical subject (the object "dial of the measuring instrument") and a predicate (the value displayed by this dial). Projecting this constraint of classicity from the mesoscopic domain onto the microscopic domain, Bohr emphasizes that each characterization relating to the microscopic domain is not that of a microscopic object, but that of a semi-classical interpretation of such object, supported by a classical interpretation of the measuring process.

It then becomes possible to account for both the mutual exclusivity of microscopic characterizations and their joint necessity, that are the two definitional components of Bohr's concept of "complementarity". The mutual exclusivity of the microscopic characterizations is explained by their "holistic" indissociability with the corresponding experimental procedures, and by the incompatibility of these experimental procedures. The joint need for several incompatible characterizations is justified by the desire to continue to consider that certain experiences, including when they are incompatible, "relate to" the same object; that is to say, by the wish to admit that, even if they do not reveal what an object is in itself, the experimental characterizations offer sketches of it.

Inspired by Höffding, Bohr considered the complementarity of characterizations in the human sciences as the expression of a similar tension. On the one hand, the mental and social processes are so integrated, so participatory, so "holistic", that one can manifest one of their aspects only by cutting them off from conditions that would manifest another aspect. On the other hand, if, in order to conform to the norms of the theory of knowledge, we still wish to subdivide these integrated-holistic processes so as to act as if an object were distinguishable from its subject, we come to the conclusion that this artificially posited object, having only "symbolic" consistency, can only be approached through several of these mutually exclusive aspects. One of Bohr's best-known examples of complementary structures is introspective psychology (just as Höffding). Better still, Bohr does not present introspective psychology as a mere illustration, but as the paradigmatic description of the epistemological limitation facing modern physics. In this example, the subject, being at the same time the object of its self-examination, must use several mutually exclusive approaches to explore herself. Thus, when the subject wants to analyze her own use of a concept, she must adopt a posture which excludes the unreflected application of this concept. The subject cannot be simultaneously actor and spectator of her own conceptual elaboration. Bohr also gave examples of complementary structures in the social sciences. For example, he wrote, benevolence is complementary to justice, since the first point of view requires personal commitment while the second point of view is a social norm.

In his fundamental philosophical text written in 1942, Heisenberg [17] transformed Bohr's thesis into a general theory of concrete human knowledge. According to Heisenberg, there exist at least three regions of knowledge that are distinguished by the degree of dissociation which can be achieved between the process of knowledge and its object.

- The first region is such that the states of things are completely separable, by a technique of search for invariants, from the process of their study. This is the region of classical physics and chemistry.
- The second region corresponds to the case where the states of things that one seeks to characterize are fundamentally inseparable from the approach adopted for the characterization. This is the region of quantum physics. But this is also, according to Heisenberg, the region of psychology and biology. Psychology is indeed marked by the fact that "(...) an essential part of what happens in the soul escapes any objective fixation because the act of fixation intervenes itself decisively in the processes". In other words, it is impossible to dissociate an object "soul" from the very act that the "soul" accomplishes. As for biology, it must take into account the incompatibility of analytical

approaches (such as physicochemical examination of cellular components), and global approaches, such as those concerned with homeostasis, behavior and intentions. If these two approaches are incompatible, then it becomes absurd to pretend to absorb one of the objects into the other, as reductionism seeks to do by excluding the categories of homeostasis and intentionality in favor of molecular concepts alone.

• The third region, finally, is one where one is interested in symbols capable of guiding not only the process of knowledge, but also life in general. This is the region of art with its formative figurations, of religion with its representations and archetypal narratives, but also of the many social practices in which an institutional symbol (fiduciary money, constitution, bill of rights) is treated as if it were an autonomous reality.

At a higher level of reflection, however, the very existence of a multiplicity of regions of knowledge is sufficient to express the impossibility to separate knowledge from the processes it uses. Heisenberg thus pointed out that "It may not be possible, in a complete description of the connections of a region, to disregard the fact that we ourselves are part of these connections." It is true that the examples of this impossibility of ignoring our commitment in the (law-like) connections which constitute a region of knowledge are found by Heisenberg in atomic physics and psychology (in his second region of knowledge). But the wording of his sentence implies than our commitment can no longer be ignored when we seek the completeness of any description whatsoever. Our commitment becomes obvious as soon as we strive to reach the boundaries of a region of knowledge, no matter which region we explore.

Many historical cases illustrate this self-revealing power of the exploration of confines. Thus, a search for completeness of the classical description of atoms and light, at the turn of the nineteenth century and the twentieth century, has led us to run up against a fundamental limitation of objectification (in the sense of complete "detachment" of an object with intrinsic properties): the limitation manifested in quantum theory. Earlier in history, a search for completeness in classical science, a science elaborated from the point of view of a detached spectator, made us stumble on the enigma of free will. As suggested by Kant, this enigma that is formulated from the point of view of the spectator could be sorted out only from the point of view of the agent.

This repeated figure of human beings meeting their own limitations at the end of a quest of knowledge has been expressed by Eddington in a celebrated sentence: "We have found a strange footprint on the shores of the unknown. We have devised profound theories, one after another, to account for its origins. At last, we have succeeded in reconstructing the creature that made the footprint. And lo! It is our own" [10].

Heisenberg made another suggestion that may help the development of a detailed relationship between quantum mechanics and the human sciences. His additional suggestion is that the impossibility of completely ignoring our own contribution can be recognized in the mathematical formalism of quantum physics itself. To understand how this is possible, we must first clarify the process by which the search for objectification overcomes obstacles. This is usually done by incorporating this obstacle into the newly defined objects and turning it into an advantage. As Heisenberg pointed out, "(...) even if a state of affairs can not be objectified in the standard sense, it remains that this fact itself can (...) be objectified in its turn and explored in its connection with other facts" [17]. In other terms, the limits of objectification are (reflectively) objectifiable. This remark illuminates the status of the mathematical formalism of quantum theory. The mathematical formalism of quantum theory is a second-order objectification of the impossibility of any first-order objectification. The first-order objectification consists in extracting spatio-temporal invariants such as classical material corpuscles with a trajectory in space and in time. It thus directly organizes the *individual* experimental phenomena observed in the space and time of the laboratory. But the second-order objectification is elaborated out of the *statistical distributions* of these experimental phenomena. It extracts an invariant predictive tool such as the state vector in a Hilbert space, thus organizing phenomena indirectly.

What is connected through the basic law of quantum mechanics, namely the Schrödinger equation, is then the state vector, rather than the individual experimental phenomena. This further expresses: (i) the impossibility of a first-order objectification (that is to say the impossibility to detach experimental phenomena from their experimental context), and (ii) the effectiveness of the second-order objectification of an invariant generator of statistical distributions. We thus realize that a considerable part of the quantum formalism has a deep meaning that extends far beyond microphysics. This formalism actually expresses a universal general epistemological situation: a stepping up of objectification in the face of the fundamental obstacles that it encounters.

Subsequently, several philosophers have taken advantage of these pioneering analyzes of the creators of quantum mechanics and elaborated a new conception of knowledge on that basis. This is the case of Karl-Otto Apel, who devoted a book to this endeavor. Starting from the two differences between natural sciences and human sciences mentioned by Habermas, Apel shows that they vanish in the quantum paradigm. The first opposition, between the (natural) sciences of detachment and the (human) sciences of commitment, collapses from the outset: "In the [sciences of nature as in the human sciences], writes Apel, it is necessary to give up the representation of an objective external world of which a multiplicity of perspectives falls, in principle, under (...) theoretical control. Instead, there are aspects of the world that are incompatible, complementary, [because they are indissolubly linked to each perspective, to each mode of intervention]" [1]. The aspects of nature seen in microphysics, "(...) are objectively incompatible and, for this reason, comparable to the mutually exclusive worldviews of the (human) sciences (*Geisteswissenschaften*)".

The second epistemological difference mentioned by Habermas between the natural sciences (supposed to describe actual properties) and the human sciences (supposed to focus on potentialities) is also analyzed and criticized by Apel.

Indeed, this difference was canceled out by the advent of quantum physics. Just as the human sciences, quantum theory manipulates symbols (the state vectors) that describe a potentiality rather than some actual event. It serves to anticipate what can happen in the future in various experimental contexts, rather than context-independent present findings. Apel attempts to identify the reason for this shift from classical physics to quantum mechanics: "(Quantum mechanics), he writes, succeeds in separating the subject from the object in the statistical explanation of the behavior of sets of particles, but it fails at the level of the individual particles". In other words, the use of probabilities in quantum physics is the mark left on the limit of first-order objectification, and the expression a second-order objectification: the indirect objectification of statistical distributions of spatio-temporal phenomena, rather than the direct objectification of a set of spatio-temporal entities.

This being granted, the modalities of the connection between quantum physics and the human sciences are of two quite distinct types. On the one hand, we can qualitatively develop the similarity of the epistemological situation between microscopic physics and each human science considered separately. On the other hand, we can seek to state the quantitative or at least formal consequences of this type of epistemological situations.

4   Qualitative Parallels Between Quantum Theory and the Human Sciences

As we have just seen, qualitative parallels between quantum theory and the human sciences can be based on the Bohrian concept of complementarity. However, complementarity being a "broad-spectrum concept" (Putnam), its modalities of applications can vary a lot. Already, in quantum mechanics, several variants of this concept have been listed by Bohr.

Taken in the broadest sense, Bohr's complementarity expresses the impossibility of getting rid of the holistic features of experimentation. But these holistic features manifest in three different ways:

- The mutual exclusivity of two variables that are inseparable from experimental contexts. An exemple of such pairs of conjugate variables is position and momentum.
- The mutual exclusivity of two pictures, respectively associated with partial experimental contexts and a global experimental context. This is the case of the corpuscular and wave pictures in the double slit experiment. The corpuscular picture is associated with the partial context of detection of the passage of an object through one slit ; and the wave picture is associated with the global context of indistinguishability of the paths corresponding to the two slits.
- The mutual exclusivity of potentiality and actuality. They correspond respectively to: (i) the context of forecasting future measurement results after the initial preparation, and (ii) the context of the final measurement. This latter

pair is represented in quantum theory by the continuous and discontinuous modes of evolution of the state vector. The "causal" mode of evolution by the Schrödinger equation excludes the "acausal" mode of evolution associated with an act of experimental localization in space-time.

Two examples of qualitative use of the Bohrian concept of complementarity in the human sciences will now be developed. They manifest the same amplitude of variation around the common theme of the holistic character of knowledge, as in quantum physics.

Klaus Meyer-Abich [21] developed a psychological variety of complementarity, in the spirit of Bohr and Höffding. According to Meyer-Abich, this kind of complementarity reflects an incompatibility between the act of aiming at objects and the act of reflecting on objectification. This incompatibility was especially highlighted by Kurt Goldstein in his empathic observations of patients who suffered brain damage during the First World War. For such patients, the psychical attitudes of intentionality and reflexivity are so dissociated that even their succession becomes impossible. Surprising as it may seem, this alteration observed by Goldstein is not a consequence of certain focused lesions, but is found in virtually all patients with extensive lesions of the cerebral cortex. In every patient of this kind, "everything that forces him to go beyond the sphere of 'actual reality' to reach what is 'simply possible', brings a failure" [12]. Patients adhere to what is immediately experienced, without being able to distantiate from it and without being capable of embedding it into a representation. They remain bound to intentional directedness towards objects without being able to step back and acquire a reflective knowledge of themselves. In a later reflection, Goldstein characterizes this deficit of patients having cortical lesions with words that were also used by Bohr: patients "act in the world instead of thinking of it or talking about it"; in other words, they become pure actors because they have lost the degrees of freedom that would have allowed them to behave as spectators of themselves as well. From this, one may infer that in organisms, the actor-stance, the stance of self-adherence to oneself, is fundamental. By contrast, the complementary stance of a detached spectator: (a) can only be incomplete, and (b) requires resources in excess of that of the actor. Holistic integration imposes the actor-stance. It allows only incidentally and fragmentarily the stance of a detached spectator, which supposes that one suspends for a time the fundamental actor-stance.

The analogy between this cognitive pathology and the complementarity of corpuscular and wave representations is striking. In the latter type of complementarity, the corpuscular representation prevails in the context of a local detection on one branch of the interferometer, whereas the wave representation is relevant in the context of an evaluation of the effects produced by the interferometer as a whole. The corpuscular representation prevails when only one path is available, while the wave representation prevails when all possible paths have been left open. Similarly, Goldstein's pathological configuration highlights a type of complementarity wherein the local approach of an act is exclusive of a global approach of action. It also corresponds to a duality of attitudes between adherence to a certain act, and the reflective distancing that allows every possibility of acting to be displayed before the eyes.

A different variety of complementarity, that still manifests the holistic nature of knowledge, is mobilized by Michael Rasmussen's reflection on linguistics [24]. Rasmussen confronts Bohr with another great Danish thinker who is almost contemporary with him: the linguist Louis Hjelmslev. However, his comparison of linguistics with the epistemological configuration of quantum physics is not limited to one author; it extends to any structuralist conception of language, such as Saussure's. The comparison is established in two steps.

The first step is a definition of observation: observing means restricting the initial conditions on the basis of which a prediction can be made. The question of the impact of the observation on the observed domain can then be replaced with a request concerning the impact of the forecast on the forecasted events. But, says Rasmussen, in linguistics, this impact is by construction considerable. When a linguist tries to predict the future evolution of her own language, she modifies it by her very act of forecasting. For, as a speaker of her language, the linguist is bound to give a normative or prescriptive value to her prediction. When she foresees the future state of her language, she prescribes a condition of identity (this future language must still be "English", despite all its transformation). And since her speech is guided by such prescription, she influences the evolution of her own language. To sum up, the prediction "disturbs" the language. The work of the linguist influences the evolutionary dynamics of her own language. Linguistic analysis cannot be detached from the metabolism of language.

The second step in this parallel between quantum mechanics and linguistic analysis consists in describing a form of complementarity. There are two mutually exclusive approaches to language: the synchronic approach and the diachronic approach. The synchronic approach tends to immobilize the language in its present form, namely in a present system of differences between signs. The diachronic approach, instead, follows in the short term the developments of the practice of speech, and it tends to identify in the long term the drifts of the system of semantic differences. Clearly, extracting a synchronic structure (from a snapshot of language), and making a diachronic analysis (of the history of language), are mutually exclusive operations. We find in Louis Hjelmslev's work a detailed description of this difference [20]. As a preliminary, Hjelmslev points out that while signs follow each other in speech, they coexist in the text that transcribes it. Their succession constitutes speech, and their coexistence constitutes texts. Both speech and texts involve a conjunction of signs. In both cases one sign comes, *and* then another, *and* then another etc. But this conjunction unfolds in time for speech, and it unfolds in space for texts.

Now, according to structuralism, each sign has a unique position which defines it by its differences with respect to the others. No other sign can *really* replace it, since to take its place would be tantamount to endorse the same pattern of differences and therefore to identify with its unique function. In this case, the signs form a disjunction: at each given position, one sign may be used *or* another, *or* another one etc. ; and this substitution changes nothing since what counts is the local pattern of differences. Hjelmslev called "relation" the conjunction typical of the textual process, and "correlation" the disjunction typical of the whole system of language. The relations between the signs of a text acquire meaning according to their positions in the system of correlations which constitutes language.

At this point, the difference in the modes of observational access becomes quite obvious. The observational access to a text is immediate, since it just consists in reading a sequence of signs. But access to the system of language is quite complex: it arises from the analysis of an immense (*a priori* unlimited) corpus of speeches and texts. Yet both types of access have a predictive value. Understanding of a text makes it possible to foresee to a certain extent what follows it, by constraining the field of future possibilities. As for knowledge of the system of language, it constrains any text and speech to fit with "grammatical" rules.

The reason for the "complementary" nature of synchronic and diachronic approaches to language is easily identifiable from there. The observation of a speech or a text forces us to accept a certain creative freedom, and consequently opens the way to a future destruction of the system that presently constrains it. The observation of the language system, on the other hand, makes it necessary to declare outlaw any deviation from it, and to set strict boundaries of what can be said without a time limit.

The most tempting analogy, although arguably a partial one, is with the third variety of Bohrian complementarity: the complementarity between actuality and potentiality. Here, the actuality is that of the text, while the potentiality is that of the system of the language, capable of generating all the texts that follow its rules and also capable of carrying a sentence of banishment against the texts which deviate from such rules.

5   Early Quantitative Applications of Quantum Theory to the Human Sciences

We must now examine the quantitative aspects of the epistemological analogy between quantum theory and the human sciences. Giving this analogy a formal translation is a decisive test for its relevance. The most delicate question for researchers in this field was how to collect in a formalism the constraints of the common epistemological configuration, while leaving aside the peculiarities of the various domains to which it extends. Let us admit, as we have said before, that the quantum formalism translates above all the limits of the activity of objectification. Does this mean that every symbol of the quantum formalism can be related to this epistemological constraint? And should we infer that the quantum theory formulated by physicists from 1925 can be transposed immediately to a number of problems of psychology, sociology, or economics? The answer to these questions is "no". Indeed, several features of the quantum formalism are derived from specific domains of the physical science, from mechanics to electrodynamics. An exemple is the structure of the Hamiltonian operator in the Schrödinger equation: its form is identical to that of the Hamiltonian function of classical mechanics, and it unambiguously expresses the connection of quantum formalism with this domain of physics. One must therefore go further up the scale of generality, and identify the epistemological core of quantum formalism after having put aside its "physical" envelope. Where do we find this nucleus: in the probabilistic algorithm of quantum theory, in its Hilbert space structure, or in the underlying structure of "orthocomplemented lattices", which was extracted by Birkhoff and Von Neumann and formulated as "quantum logic"? Each of these options has been explored (albeit sporadically) during the twentieth century. I will mention three of them.

According to Jean-Louis Destouches (1909–1980), it is the structure of the probabilistic algorithm that expresses what quantum formalism owes to the epistemological situation confronting microscopic physics [8]. Destouches thus tried to build what he called a "general theory of predictions" capable of providing probabilistic evaluations ; and he identified in it the particular features that make it possible to arrive at quantum or classical versions of this kind of theory. During his research, worked out jointly with Paulette Destouches-Février, he obtained an important result that retrospectively justified his initial program. This result is stated in the following theorem: "If a theory (of predictions) is objectivist, it is in principle deterministic and one can thus define an intrinsic state of the observed system ... Conversely, if a theory (of of predictions) cannot be considered as objectivist, that is to say if it is irreducibly subjectivist, then it is not deterministic in principle; as a consequence, such theory is essentially indeterministic" [9]. In the latter case, the theory is bound to be probabilistic, whereas in the former case the use of probabilities is only due to the ignorance of the intrinsic state of the system. The result is remarkable, but the sentences by which it is expressed involve a vocabulary that may trigger misunderstandings. To begin with, the couple of terms "objectivist-subjectivist" expresses the opposition between an epistemological situation where the work of objectification can be carried out, that is to say, where it leads to object-specific autonomous determinations, and another situation (typical of quantum physics) where the phenomena are inseparable from the instrumental context that allows them to manifest. This translation of the term "subjectivist", with its inappropriate connotations, by a more neutral term as "contextualist", is justified by the definition given by Paulette Destouches-Fevrier herself: "(We call) 'objectivist theory' a theory in which measurement results can be considered as intrinsic properties of observed systems, and 'subjectivist theory' a theory in which the measurement results cannot be ascribed to the observed system as intrinsic properties, but only to the complex apparatus-system, with no possible analysis that would ascribe part of the result to each one (...)". Taking into account these definitions, the heart of the theorem can be stated as follows: a theory allowing to predict phenomena indissociable from their mode of access is "essentially indeterministic". The ineliminable use of probabilities is therefore, according to Jean-Louis Destouches and Paulette Destouches-Février, the generic mark of the epistemological situation of microscopic physics.

That being granted, is there a way to identify, among the features of the probabilistic formalism of quantum theory, something that is specifically expressive of this epistemological situation, after having set aside what belongs to the physical domain to which it applies? Doing that is precisely one of the goals that Destouches set for himself when he developed his general theory of previsions in the late 1930s. This general theory of previsions, he writes, makes it possible to "... separate hypotheses about predictions from the strictly physical assumptions" [7].

As a first step, Destouches defines the initial "prediction element" that is characteristic of a given experiment. A "prediction element" is a mathematical entity that can be used to associate a probability distribution to each measurement that can be made after some given experimental preparation.

The second step consists in calculating the evolution of the prediction element in time. This is done by using a unitary operator, which has the property of ensuring that the sum of probabilities evaluated from the prediction element remains equal to 1 at any time.

The third step consists in making a list of "eigen (or proper)" prediction elements, which provide a probability 1 for one of the values that the selected variable can take, and 0 for all other values of this variable.

At the fourth step, one determines the set of coefficients such that the final prediction element can be written as a linear combination of the proper prediction elements, weighted by these coefficients (we thus generate a vector space of prediction elements that may, if certain additional conditions are met, acquire the structure of a Hilbert space).

At the fifth step, finally, the probability of each value of the measured variable is calculated. This last stage is especially interesting because from it, one may bring out a characteristic imprint, on the form of the probabilistic evaluations, of the epistemological situation of inseparability of the phenomena vis-à-vis their experimental modes of access. When the predictions concern such inseparable phenomena, one can prove a theorem stating that the probabilities are the square modulus of the previous coefficients. This is the "Born rule", which generates probability distributions that are isomorphic to the intensities of a wave. Through this theorem demonstrated by Paulette Destouches-Février, Born's rule and the wave-like effects typical of quantum mechanics have both been shown to be direct consequences of the limit to objectification that characterizes microscopic physics.

Now, we are certain that there is indeed a feature of the quantum predictive formalism (the Born rule) which directly expresses the epistemological situation of indissociability of phenomena with respect to their modes of access. But what about other features ? What in the structure of the general theory of predictions is still connected to physics? Two things, essentially: (a) the definition of each variable, because it depends on the procedure used for its measurement; and (b) the structure of the unitary operator that is used to calculate the evolution of the prediction elements, because it expresses the dynamics of the process under consideration. In standard quantum mechanics, this evolution operator is inserted in the Schrödinger equation ; it involves a Hamiltonian operator derived from classical mechanics or electrodynamics. All the rest of the predictive formalism (including the vector space structure) is much more general than any physical theory. A momentous consequence of this generality was drawn by Destouches in the 1950s: the quantum-like theory of predictions applies to "many other domains" than physics. In particular, it was applied by Destouches to biology and to "questions of econometrics" [7], thus showing its relevance for some human sciences.

Two other authors (Satosi Watanabe and Patrick Heelan) sought the similarity between quantum physics and the human sciences in an even deeper structure, underlying the probabilistic formalism. They found it in the "ortho-complemented lattice algebra", which replaces in quantum theory the ordinary Boolean algebra of the empirical propositions of classical science. This structure is at once looser and more general than that of Boolean algebras; it can be considered as a non-Boolean network of Boolean subalgebras. As Watanabe pointed out [28], the use of an orthocomplemented lattice algebra instead of a Boolean algebra is a mark of a deep alteration of the epistemological situation. Indeed, the Boolean algebra of empirical propositions is underpinned by a postulate according to which "each predicate corresponds bi-univocally to a defined set of objects that satisfy the predicate". In other words, Boolean algebras apply to a corpus of propositions which define subsets of objects characterized by the intrinsic possession of a predicate. Things become very different when a measurement result can no longer be assigned to an object as its intrinsic attribute. If this is the case, if we must suspend the attribution of predicates to objects, if we cannot even set apart "primary qualities" belonging to objects from "secondary qualities (or predicates)" relating to experimental methods, then the former postulate is no longer valid, and Boolean algebra no longer governs all empirical propositions. What comes in the place of Boolean algebras is a non-distributive orthocomplemented lattice algebra which articulates Boolean subalgebras within a structure that is more universal than the latter.

Watanabe ascribes these results a generality that far exceeds physics alone. In order to test their generality, he applies them to the composite structure of everyday language. This language, he points out, combines effortlessly mentalist and physicalist elements in the same propositions. The option one adopts regarding the legitimacy or illegitimacy of such a combination partly determines the position one occupies in the debate on the mind-body problem. Considering that the mentalist predicates (about 'inner' states) should not be combined in the same sentence with physicalistic predicates (on the states of the body), but that both are legitimate, is to engage on a path that leads to dualism. Giving priority to physicalistic predicates (considering mentalist predicates as redundant) is to engage in the path of reductionism or even eliminativism. It remains to be seen what are the conditions of possibility of the curious association of the two kinds of predicates which is so common in ordinary language. Watanabe begins with emphasizing that this association is by no means obvious. The famous remark made by Gilbert Ryle, according to which the mentalist predicates are dispositional in nature, whereas the standard physicalist predicates are of a categorical nature, makes their juxtaposition in the same sentence almost baroque. This difference in nature also renders a strict correlation between mentalistic and physicalist predicates implausible: (a) a pure disposition may admit to being conditioned by a fairly wide range of physical states, and (b) information derived from complex mental states are probabilistic, whereas the physicalist propositions are of the assertoric type. In addition, the modes of access to the two types of predicates are profoundly different, not to say incompatible. Access to macroscopic physicalist predicates takes place through a single observation or measurement, whereas access to the dispositional traits of mentalist predicates can only be achieved through the study of an open-ended sequence of behaviors. Besides (as Bohr pointed out), access to the entirety of the alleged physical substratum of a dispositional mental predicate would destroy this substratum, and would thus be eminently disturbing for the mental state. The two languages, mentalist and physicalist, thus prove to be mutually exclusive in a sense very similar to that of the conjugated variables of quantum mechanics.

Therefore, if we want to understand how ordinary language is able to combine mentalist and physicalist propositions in one and the same discourse, it is necessary to aknowledge that this language can be underpinned by a nonclassical logic. The latter logic is shown to be isomorphic to quantum logic, i.e. to a non-distributive orthocomplemented lattice. This is enough to see the mind-body problem under a new light. The mind-body problem stems from the wrong attempt to project on a single Boolean logic two classes of propositions and predicates that are mutually exclusive due to the incompatibility of the corresponding modes of access. The mind-body problem is then dissolved when one has accepted to take this duality of modes of access into account (yet without hypostatizing them into property dualism).

Patrick Heelan later extended Watanabe's cogent analysis to any context-dependent language that one purports to unify through a common meta- or trans-contextual language [16]. To illustrate this extension, Heelan applied the former analysis in ethno-linguistics. According to him, a meta-contextual language (namely a language that can be used by speakers of two linguistic subgroups who attempt to communicate with each other) is necessarily underpinned by a logic that is isomorphic to the quantum logic of Birkhoff and Von Neumann.

6 Epilogue

Perhaps, however, the most interesting lesson that can be drawn from these applications of quantum theory to the human sciences does not concern the latter sciences, but physics itself. In view of the successes obtained in perceptual psychology or in decision theory by applying a proto- or quasi-quantum formalism, it becomes difficult to deny the epistemological-reflective meaning of quantum mechanics. For the only plausible common feature of psychology, sociology, economics, decision-making, and quantum mechanics is the type of act of knowledge that these disciplines bring into play. Bohr's provocative statement (according to which "physics is to be regarded not so much as the study of something a priori given, but rather as the development of methods of ordering and surveying human experience" [5]) can now avail itself of its concretization in a multiplicity of domains whose isomorphism reveals a common epistemological approach towards what is still called, by habit, their "object".

The fact that quantum theory apply so well to many human sciences then lends credibility to recent non-representationalist and thoroughly probabilist interpretations of quantum mechanics. Two of them, that were already mentioned in the introduction, are Richard Healey's "pragmatic" interpretation and Christopher Fuchs' "QBism".

According to Richard Healey [15], quantum theory is a new kind of non-representational science ; it is fundamental but with no "beables" in it ; it is objective but its kind of objectivity consists in universal prescriptions to agents ; its symbols are purely predictive, not descriptive (therefore, famous features such as entanglement do not express a "real" entanglement of physical systems).

As for Christopher Fuchs [11], he considers that the symbols of quantum mechanics, such as state vectors, represent nothing more than a mathematical instrument used by agents in order to make optimal bets about the outcomes of future experiments. In Fuchs' own terms, the quantum symbolism is just a "user's manual" for each individual agent. If one wants to "explain" a phenomenon by quantum mechanics, the very meaning of the word explanation has to be changed: quantum mechanics explains why, in order to be coherent, an agent should assign probability p, not why a certain result has been obtained. Accordingly, standard paradoxes are dissolved by removing entirely the ontological import of symbols. A crucial example is the Einstein Podolsky Rosen paradox, that is dissolved by considering that predicting a future outcome with probability 1 does not imply the existence of Einstein's "element of reality", but only expresses the supreme confidence of agents.

In the wake of these deflationary interpretations of quantum mechanics, even the elementary presuppositions that measurement results are about "physical systems" has been challenged. The analysis of sequences of measurement outcomes, in the framework of the so-called "device-independent approaches" [13], has shown that their structure is generally incompatible with the concept of "permanent physical systems bearing properties". This puts an end to the implicit but pervasive idea that quantum mechanics describes the exceptional features of certain microphysical systems ; it rather reveals the collapse of standard ontological patterns at the micro-scale, and the emergence of a context-dependent kind of knowledge [2–4]. This is enough to give full legitimacy to a transposition of quantum theory to the human sciences. For, even though there can be no common ontological domain between microphysics and the human sciences, there is a common epistemological approach that determines the structure and meaning of both disciplines.

References

1. Apel, K.O.: L'a priori du corps dans le problème de la connaissance. Cerf (2005)
2. Bitbol, M.: Mécanique quantique, une introduction philosophique. Flammarion (1996)
3. Bitbol, M.: Some steps towards a transcendental deduction of quantum mechanics. Philosophia Nat. **35**, 253–280 (1998)
4. Bitbol, M.: De l'intérieur du monde: pour une philosophie et une science des relations. Flammarion (2011)
5. Bohr, N.: The Unity of Human Knowledge, Philosophical Writings of Niels Bohr, vol. 3. Ox Bow Press, Woodbridge (1960)
6. Cavaillès, J.: Du collectif au pari, àpropos de quelques théories récentes sur les probabilités. Revue de Métaphysique et de Morale **47**(2), 139–163 (1940)
7. Destouches, J.L.: Prévisions, calcul et réalités. Les Grands problèmes des sciences (1965)
8. Destouches, J.L.: La Mécanique Ondulatoire. Les Etudes Philosophiques **4**(3), 473–474 (1949)
9. Destouches-Février, P.: La structure des théories physiques. Presses Universitaires de France (1951)
10. Eddington, A.S.: Space, Time and Gravitation. Cambridge University Press, Cambridge (1920)
11. Fuchs, C.A.: QBism, the perimeter of quantum Bayesianism. arXiv preprint arXiv:1003.5209 (2010)
12. Goldstein, K.: La structure de l'organisme. Gallimard, Paris (1951)
13. Grinbaum, A.: How device-independent approaches change the meaning of physical theory. Stud. Hist. Philos. Sci. Part B: Stud. Hist. Philos. Mod. Phys. **58**, 22–30 (2017)
14. Habermas, J.: Logique des sciences sociales et autres essais. Presses Universitaires de France, Paris (1987)
15. Healey, R.: The Quantum Revolution in Philosophy. Oxford University Press, Oxford (2017)
16. Heelan, P.A.: Complementarity, context dependence, and quantum logic. Found. Phys. **1**(2), 95–110 (1970). https://doi.org/10.1007/BF00708721
17. Heisenberg, W.: Philosophie, Manuscrit de 1942. Edition du Seuil (1998)
18. Høffding, H.: Filosofiske Problemer. Univ. Bogtr. (1902)
19. Høffding, H.: Relation som Kategori. Kluwer, Dordrecht (1921). Quoted by J. Faye, Niels Bohr, his Heritage and Legacy
20. Hjelmslev, L.: Prolégomènes àune théorie du langage. Minuit (1971)
21. Meyer-Abich, K.: Bohr's Complementarity and Goldstein's holism in reflective pragmatism. Mind Matter **2**, 91–103 (2004)
22. Petersen, A.: The philosophy of Niels Bohr. Bull. At. Sci. **19**(7), 8–14 (1963). https://doi.org/10.1080/00963402.1963.11454520
23. Piaget, J.: Logique et connaissance scientifique. Gallimard-Pléiade, Paris (1967)
24. Rasmussen, M.: Le problème de l'observation en linguistique. Une comparaison entre Louis Hjelmslev et Niels Bohr. Louis Hjelmslev et la sémiotique contemporaine **24**, 112 (1993)
25. Ryle, G.: The Concept of Mind. The University of Chicago Press, Chicago (1949)
26. Searle, J.: The Construction of Social Reality. Allen Lane, Bristol (1995)
27. Wang, H., Sun, Y.: On quantum models of the human mind. Top. Cogn. Sci. **6**(1), 98–103 (2014)
28. Watanabe, S.: Algebra of observation. Prog. Theor. Phys. Suppl. **37**, 350–367 (1966)

Quantum Cognition

The Power of Distraction: An Experimental Test of Quantum Persuasion

Ariane Lambert-Mogiliansky[1(✉)], Adrian Calmettes[2], and Hervé Gonay[3]

[1] Paris School of Economics, 48 boulevard Jourdan, 75014 Paris, France
alambert@pse.ens.fr
[2] Department of Political Science, The Ohio State University, 2140 Derby Hall, 154 N Oval Mall, Columbus, OH 43210, USA
calmettes.1@osu.edu
[3] GetQuanty, 54 rue Greneta, 75002 Paris, France
herve.gonay@getquanty.com

Abstract. Quantum-like decision theory is by now a well-developed field. We here test the predictions of an application of this approach to persuasion as developed by Danilov and Lambert-Mogiliansky in [6]. One remarkable result entails that in contrast to Bayesian predictions, distraction rather than relevant information has a powerful potential to influence decision-making. We conducted an experiment in the context of donations to NGOs active in the protection of endangered species.

We first tested the quantum incompatibility of two perspectives 'trust' and 'urgency' in a separate experiment. We next recruited 1371 respondents and divided them into three groups: a control group, a first treatment group and the main treatment group. Our main result is that 'distracting' information significantly affected decision-making: it induced a switch in respondents' choice as to which project to support compared with the control group. The first treatment group which was provided with compatible information exhibited no difference compared with the control group. Population variables play no role suggesting that quantum-like indeterminacy may indeed be a basic regularity of the mind. We thus find support for the thesis that the manipulability of people's decision-making is linked to the quantum indeterminacy of their subjective representations (mental pictures) of the choice alternatives.

Keywords: Persuasion · Distraction · Information processing · Belief updating · Quantum cognition

1 Introduction

The theory of persuasion was initiated by Kamenica and Gentzkow [12] and further developed in a variety of directions. The subject matter of the theory of persuasion is the use of an information structure (or measurement) that generates new information in order to modify a person's state of beliefs with the intent

© Springer Nature Switzerland AG 2019
B. Coecke and A. Lambert-Mogiliansky (Eds.): QI 2018, LNCS 11690, pp. 25–38, 2019.
https://doi.org/10.1007/978-3-030-35895-2_2

of making her act in a specific way. The question of interest is how much can a person, call him Sender, influence another one, call her Receiver, by selecting a suitable measurement and revealing its outcome. An example is in lobbying. A pharmaceutical company commissions to a scientific laboratory a specific study of a drug impact, the result of which is delivered to the politician. The question of interest from a persuasion point of view is what kind of study best serves the company's interest.

Receiver's decision to act depends on her beliefs about the world. In [12] and related works, the beliefs are given as a probability distribution over a set of states of the world. A central assumption is that uncertainty is formulated in the standard classical framework. As a consequence the updating of Receiver's beliefs follows Bayes' rule.

However as amply documented the functioning of the mind is more complex and people often do not follow Bayes rule. Cognitive sciences propose alternatives to Bayesianism. One avenue of research within cognitive sciences appeals to the formalism of quantum mechanics (QM). A main reason is that QM has properties that reminds of the paradoxical phenomena exhibited in human cognition. As argued by Danilov and Lambert-Mogiliansky in [6], there also exists deeper reasons for turning to quantum mechanics when studying human behavior. Moreover cognition has been successful in explaining a wide variety of behavioral phenomena such as disjunction effect, cognitive dissonance, order effects or preference reversal (see [3,10]). Finally, there exists by now a fully developed decision theory relying on the principle of quantum cognition (see [7]). Clearly, the mind is likely to be even more complex than a quantum system but our view is that the quantum cognitive approach already delivers interesting new insights in particular with respect to persuasion.

In quantum cognition, the system of interest is the decision-maker's mental representation of the world. It is represented by a *cognitive state*. This representation of the world is modelled as a quantum-like system so the decision relevant uncertainty is of non-classical (quantum) nature. This modelling approach allows capturing widespread cognitive difficulties that people exhibit when constructing a mental representation of a 'complex' alternative (cf, [4]). The key quantum property that we use is that some characteristics (cf. properties) of a complex mental object may be "Bohr complementary" that is incompatible in the decision-maker's mind: she cannot combine in a stable way pieces of information from the two perspectives. A central implication is that measurements (new information) modifies the cognitive state in a non-Bayesian well-defined manner.

As in the classical context our rational Receiver uses new information to update her beliefs so that choices based on updated preferences are consistent with ex-ante preferences defined for the condition (event) that triggered updating. In [7], we learned that a dynamically consistent rational quantum-like decision-maker updates her beliefs according to the von Neumann-Lüders postulate. In two recent papers, important theoretical results were established. First, as shown in [5], in the absence of any constraints on measurements, full persuasion applies: Sender can always persuade Receiver to believe anything that favors him. Next, in [6] the same authors investigate a short sequence of measurements but in the frame of a simpler task that they call "targeting". The object of "targeting" is the transition of a belief state into another specified target state. The main results of relevance to our issue is that distraction providing 'not relevant' or 'incompatible' information has significant persuasion power. This is in sharp contrast with the Bayesian context where such information should have minimal or no effect at all.

The present paper aims at testing experimentally those predictions. More precisely, we want to test whether a question (a measurement) addressing a perspective that is incompatible with the information relevant for the decision at stake can affect decision-making. That is we test the concluding statement in Akerlof and Shiller's book (see [1]) "just change the focus of people's mind and you change the decisions they make".

In the psychology literature, the distraction effect was first introduced by Festinger and Maccoby in [9]. Its link with persuasion has now proved empirically valid through many different experimental contexts (for a review, [2]).[1] Interestingly studies (see [16]) have shown how a noninformative signal can decrease documented resistance to persuasion (see [8]). In addition, across five different experimental contexts and content domains, Kupor and Tormala revealed in [14] that interruptions that temporarily disrupt(distract) a persuasive message can increase consumers' processing of that message; consumers being more persuaded by interrupted messages than they would be by the exact same messages delivered uninterrupted.

The situation that we consider is the following. People are invited to choose between two projects aimed at saving endangered species (Elephants and Tigers). The selected project will receive a donation of 50€(one randomly selected respondent will determine the choice). We consider two perspectives of relevance for the choice: the urgency of the cause and the trustworthiness (or honesty) of the organization that manages the donations. As a first step, in a separate experiment we establish that the two perspectives are incompatible by exhibiting a significant order effect (as in [3, 17]. In the main experiment, respondents were divided into three groups: a control and two treatment groups. They all go through a presentation of the projects and some questions about their preferences. The difference between the groups is that the first treatment group receives general additional information compatible with their (elicited) preferences while the second one receives general additional information incompatible with their preferences.

The results are in accordance with the predictions of the theoretical model: incompatible information has a significant impact such that the respondents on the whole switched their choice as compared with the control. Compatible

---

[1] Decades of research on social influence have emphasized two distinct routes to persuasion: the "central" route and the "peripheral" route. According to Petty and Cacioppo in [16], the central route involves influence that takes place as a result of relatively deep processing of information that is high in message relevance, whereas the peripheral route involves influence that takes place as a result of relatively superficial processing of information that is low in message relevance.

information has globally no impact compared to the control group. None of the population variable has any impact. This suggests that the quantum model may indeed capture basic regularities of the mind relevant to decision-making.

2 The Quantum Persuasion Approach

Let us first briefly describe the classical approach (Bayesian persuasion) developed by Kamenica and Gentzkow [12]. We have a person call him Sender who tries to influence a decision-maker call her Receiver by means of an information structure or a measurement that generates information. Information affects Receiver's beliefs which in turn affect her evaluation of uncertain choice alternatives and therefore the choice she makes. In the classical context Receiver updates her beliefs using Bayes rule and therefore the power of Sender is constrained by Bayesian plausibility - that is the expected posteriors must equal the priors.

The quantum persuasion approach has been developed in the same vein as the Bayesian persuasion. A central motivation is that persuasion seems much more effective than what comes out of the Bayesian approach. So instead of assuming that agents are classical, it has been proposed that they are quantum-like. This means that the representation of reality upon which they make decision does not evolved according to Bayes rule but follows instead von Neumann-Lüder's rule (vNL). vNL updating has been shown to be an expression of dynamic consistency in such a context (see [7]).

The present paper aims at experimentally testing some predictions of the theory of quantum persuasion. More precisely as shown in [6], Sender can use 'distracting' measurements as tools to influence Receiver. A distracting measurement corresponds to a measurement that generates information that is incompatible (or Bohr complementary) with the information used by Receiver to evaluate the choice alternatives for decision-making. The objective is to switch the focus of Receiver's mind (distract her) which changes her cognitive state or her beliefs although no information relevant to her concern is provided. The following example from [5] illustrates the point.

Example

A consumer is considering the purchase of a second hand smartphone at price 30€ of uncertain value to her. What matters to her is its technical quality which may be standard or excellent. She holds beliefs about the quality of the smartphone. Based on those beliefs, she assigns an expected utility value to the smartphone which determines her decision whether or not to buy the item. Receiver's expected utility for the smartphone in belief state B is represented by the trace of the product of operators A and B[2]:

$$Eu\left(A;B\right) = \mathbf{Tr}\left(AB\right) = (1/5) * 100 + (4/5) * 0 = 20 < 30. \tag{1}$$

Given belief B Receiver does not want to buy the smartphone.

---

[2] For a complete formulation of choice theory in the quantum context, see [7].

Can Sender persuade Receiver to buy by selecting an appropriate measurement? Consider another property (perspective) of the smartphone that we refer to as Glamour (i.e. whether celebrities have this brand or not). The two properties (perspectives) are assumed incompatible in the mind of Receiver. Receiver can think in terms of either one of the two perspectives but she cannot synthesize (combine in a stable way) pieces of information from the two perspectives. This is illustrated in Fig. 1.

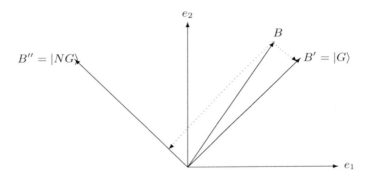

Fig. 1. Receiver's cognitive state.

Assume that Sender brings up the discussion to the Glamour perspective and performs the measurement so Receiver learns whether her preferred celebrity has this smartphone. With some probability p (say 0.9) the new cognitive state is $B' = G$ and with the complementary probability $1 - p = .1$ it is $B'' = NG$.
We note that:

$$Eu\,(A; B') = \mathbf{Tr}\,(AB') = 50 > 30. \tag{2}$$

$$Eu\,(A; B'') = \mathbf{Tr}\,(AB'') = 50 > 30. \tag{3}$$

In both cases Receiver is persuaded to buy and Sender gets a positive utility.

In the example above it is easy to show that the new belief state violates Bayesian plausibility - the expected posteriors for the event 'the smartphone is excellent' are:

$$5p + .5(1 - p) = .5 \tag{4}$$

which is larger than the priors which equals 0.2. Moreover the measurement Glamour is not relevant to the beliefs about the quality of the used smartphone. It is a distraction. Yet it affects the beliefs and the associated decision. In a Bayesian context irrelevant/uninformative data do not modify beliefs and therefore cannot be used as means of persuasion.

3   Experimental Design

Our main experiment uses the property of Bohr complementarity of mental perspectives, i.e., their possible incompatibility in the mind of Receiver. More precisely it relies on the hypothesis that two perspectives of interest are incompatible. The two perspectives that we consider are "the urgency of the task" (Urgency) and the "trustworthiness (Honesty)[3] of the organization that manages the funds". As a first step we want to provide support for this hypothesis. Two properties are incompatible if measuring them in different orders yields different results. Therefore, we started with an experiment to check whether order matters for the response profile obtained.[4]

3.1   Testing for Perspective Incompatibility

295 participants completed a short survey on the website Typeform. They were recruited through Amazon's Mechanical Turk; for which data quality has been confirmed by different studies (see [13]). They were paid $0.1 and spent on average 0:17 min to complete the survey.

Participants were first presented a short description of the situation of refugees in Myanmar with mention of the main humanitarian NGO present on the field.

"About a million refugees (a majority of women and children) escaped persecution in Myanmar. Most of them fled to Bangladesh. The Bengali Red Crescent is the primary humanitarian organization that is providing help to the Rohingyas. They are in immediate need of drinkable water, food, shelter and first medical aid."

They were then asked to evaluate the urgency of the cause and the trustworthiness of the NGO on a scale from 1 ("Not urgent" or "No trust") to 5 ("Extremely urgent" or "Full trust"). The order of presentation of each question was randomized so that half of participants responded to the urgency question before trust (U-T), and the other half conversely (T-U).

The data were processed, cleaned and analyzed with Stata. Probit regression models were used to analyze the effect of the order of the questions on the responses. In addition, because we were only interested in decision switches, responses were clustered into two groups: low level of urgency (resp. trust) (responses ≤ 3) and high level of urgency (resp. trust) (responses > 3).

The results show that the order of the question impacts significantly the responses given to both Urgency (*p-value* = .050)[5] and Trust (*p-value* = .026). This can be seen on Table 1.

There exists other (psychological) theories that account for order effects such as primacy or recency effects. Yet, the results indicate that there seems to exist

---

[3] The two terms are used interchangeably consistently with the definition given to honesty - see below.

[4] Note that even in Physics there is no theoretical argument for establishing whether two properties are compatible or not. This must be done empirically.

[5] Note that we consider significant a p-value that is exactly equal to .05.

Table 1. Regression matrix for Order effects

	(1) TRUST	(2) URGENCY
Main	−0.330*	−0.323*
ORDER	(0.026)	(0.050)

p-values in parentheses
* $= p \leq 0.05$, ** $= p \leq 0.01$,
*** $= p \leq 0.001$
Notes. ORDER $= 0$ for U-T,
ORDER $= 1$ for T-U

a primacy effect for one question, but a recency effect for the other. We thus rejected the primacy and recency explanations and in the remaining we view the two perspectives as incompatible.

3.2 Quantum Persuasion

The participants were divided into three groups. Two treatment groups and a control group as explained below. All three groups were presented a screen with an introductory message, informing them that the questionnaire was part of a research project on quantum cognition and that they will contribute in deciding which one of two NGOs projects will receive a 50€ donation. Presumably this created an incentive to respond truthfully. They were then asked to click on a button that would randomly assign them to a given condition. In all conditions, participants were shown a short text about elephants and tigers in association with an NGO working for their protection, namely the Elephant Crisis Fund (ECF) and Tiger Forever (TF).[6] The order of presentation of the text was reversed for half of the subjects. This aimed at avoiding order effects irrelevant to our point. The texts contained a brief description of the dramatic situations of elephants (resp. tigers) and of ongoing actions by the NGOs.

"Elephant crisis fund: A virulent wave of poaching is on-going with an elephant killed for its tusks every 15 min. The current population is estimated to around 700 000 elephants in the wild. Driving the killing is international ivory trade that thrives on poverty, corruption, and greed. But there is hope. The Elephant Crisis Fund closely linked to World Wildlife Fund (WWF) exists to encourage collaboration, and deliver rapid impact on the ground to stop the killing, the trafficking, and the demand for ivory."

"Tiger Forever: Tigers are illegally killed for their pelts and body parts used in traditional Asian medicines. They are also seen as threats to human communities. They suffer from large scale habitat loss due to human population

---

[6] Individually speaking, the Urgency and the Honesty perspectives could be different for a refugee problem compared to an endangered species one. However, we always compared the perspectives in light of a donation to an NGO. In addition, given the results, we thus do not consider that difference to be significant.

growth and expansion. Tiger Forever was founded 2006 with the goal of reversing global tiger decline. It is active in 17 sites with Non-Governmental Organizations (NGOs) and government partners. The sites host about 2260 tigers or 70% of the total world's tiger population".

It is worth mentioning that the descriptions were formulated so as to slightly suggest that the elephants' NGO (ECF) could be perceived as being more trustworthy (because of its link with of WWF, a well-known NGO). The text about tigers in contrast suggests a higher level of urgency (a mention was made of the absolute number of remaining tigers, significantly lower than the number of remaining elephants). Thereafter all respondents were asked:

"When considering donating money in support of a specific project to protect endangered species, different aspects may be relevant to your choice. Let us know what counts most to you". It followed:

"The urgency of the cause: among the many important issues in today's world, does the cause you consider belong to those that deserve urgent action?" or "The honesty of the organization to which you donate: do you trust the organization managing the project to be reliable; i.e. do you trust the money will be used as advertised rather than diverted."

The objective was to elicit their preferences in the sense of which perspective was most important. The rest of the experiment depended on which one of three groups the participants belonged to.

In the first control condition (baseline) they were next asked whether or not they wanted to read the first descriptions again or if they wanted to make their final decision i.e., to choose between supporting Elephant Crisis or Tiger Forever both represented by an image of an adult tiger respectively adult elephant (also delivered in random order on a line).

In the first treatment condition, they were redirected to a screen with general information compatible with the aspect they indicated as determinant to their choice when making a donation. Importantly the information did not favor or disfavor directly or indirectly any of the two projects. Rather, the Honesty extra-information dealt with NGOs' integrity *in general*, and the Urgency one with *global* mass extinction. Then they were offered the opportunity to read again the descriptions before deciding or directly make their image choice between ECF and TF. Those who responded honesty saw:

"Did you know that most Elephant and Tiger projects are run by Non-Governmental Organizations (NGOs)? But NGOs are not always honest ! NGOs operating in countries with endemic corruption face particular risks. NGOs are created by enthusiastic benevolent citizens who often lack proper competence to manage both internal and external risks. Numerous scandals have shown how even long standing NGOs had been captured by less scrupulous people to serve their own interest.

So a reasonable concern is whether Tiger Forever or Elephant Crisis Fund deserves our trust."

Those who responded urgency saw:

"Did you know that global wildlife populations have declined 58% since 1970, primarily due to habitat destruction, over-hunting and pollution. It is urgent to reverse the decline! "For the first time since the demise of the dinosaurs 65 million years ago, we face a global mass extinction of wildlife. We ignore the decline of other species at our peril – for they are the barometer that reveals our impact on the world that sustains us."—Mike Barrett, director of science and policy at WWF's UK branch. A reasonable concern is how urgent protecting tigers or elephants actually is."

In the main treatment condition, participants were redirected to a screen with general information on the aspect they did not select as determinant. So those who selected honesty (resp. urgency) saw the screen on global wildlife decline (resp. NGO's scandals). Then they were offered the opportunity to read again the description before deciding or directly make their image choice.

Finally, information about their age, gender, education and habits of donation to NGOs was collected before the thank-you message and the end of the experiment.

1371 participants completed the survey on the website Typeform and were recruited through Amazon's Mechanical Türk. They were paid either $1, $0.8 or $0.75 (for the shorter baseline survey) depending on their conditions. They spent on average 1:33 min for the experiment. 49% of them were females, 61% males, the mean age was 35, and their mean education level was undergraduate level.

3.3   Predictions

Before getting into the results and their interpretation. Let us remind of what the main predictions are.

First, the predictions of both the Bayesian and the quantum model regarding treatment group 1 (who received compatible information) are similar to the extent that general information should have a marginal or no effect at all compared with the control group's choice of whom to donate.[7] This treatment group allows rejecting the argument that any additional information upsets people's beliefs so as to significantly affect their choice.

In contrast the predictions of the two models regarding the main treatment group are distinct. The Bayesian model predicts that general information on an issue that is not determinant to choice should have no effect or a very small counter-balancing effect. That effect would be due to the fact that even if say "trustworthiness" is determinant, it needs not mean that urgency is irrelevant. In contrast, the quantum persuasion model predicts that the distraction provided to the treatment group could significantly modify the allocation of responses compared to the control group. It should be noted that since we lack information

---

[7] In a companion paper, we investigate in details the mechanisms behind those predictions. In the quantum case we do have effects due to the measurement but they tend to neutralize each other. In the classical case the effect if not null, is small and depends on the initial conditions.

about the exact correlation coefficients between the two perspectives, we do not have precise quantitative predictions.

4   Results

Data were processed, cleaned and analyzed with Stata. Mainly due to missing values, but also to solve a technical misstep[8] and in order to equally balance the number of participants in each condition, 471 participants had been removed from the data. Probit regression models were conducted to analyze the impact of the variables of interest.

4.1   Descriptive Statistics

As shown in Table 2, overall, 72,7% of the participants valued the Honesty of the NGO more than the Urgency of the cause, 87,2% made their final decision without reading the descriptions a second time, and 54,6% voted for Elephants Crisis Fund (ECF). Furthermore, after further distinctions, we observe that while most of participants preferred elephants to tigers in the control condition and in the compatible one (59% and 56% respectively), the tendency reverses for the incompatible condition (51% chose tigers). Regarding the revealed preferences, overall, the majority of participants who preferred Urgency chose Tigers (52%), whereas the majority of those who preferred Honesty chose Elephants (57%).

Table 2. Descriptive statistics

Variable	Mean	Std. Dev.
ChoiceHU	0.727	0.446
DecisionRead	0.872	0.334
FinalChoice	0.546	0.498
Age	35.368	10.522
Gender	0.606	0.489
Education	1.98	0.706
NGO	0.424	0.495

Notes. ChoiceHU-choice between Urgency (=0) and Honesty (=1); DecisionRead-decision to read the descriptions again (=0) or not (=1); FinalChoice-final choice between Tigers (=0) and Elephants (=1); Gender-females (=0) and males (=1); Education-highest level of formal education between secondary school (=0), high school (=1), undergraduate (=2), graduate and over (=3); NGO-donation of either nothing (=0) or something (=1) in the last 3 years.

---

[8] Some participants were likely to have taken the questionnaire twice and so were deleted.

4.2 Data Analysis

As Table 3 shows, the first set of results establishes that incompatible information has a statistically significant impact on the final choice (p-$value = .011$), whereas the compatible information did not significantly lead to different results compared to the baseline (p-$value = .314$). The effect of the incompatible condition on the final decision thus seems to be as expected i.e. it significantly reverses the direction of the choice observed in the control condition. More precisely, everything else being constant, the predicted probability of choosing Elephants is 10.52% (*marginal effect*) lower for an individual in the incompatible condition.

Table 3. Regression matrix for Final Choice

	(1) FinalChoice	(2) FinalChoice	(3) FinalChoice	(4) FinalChoice	(5) FinalChoice
FinalChoice					
Info	−0.0936 (0.364)	−0.100 (0.335)			−0.105 (0.314)
InfoIncomp	−0.270** (0.009)	−0.259* (0.013)			−0.265* (0.011)
Age		0.00198 (0.630)		0.00261 (0.523)	0.00171 (0.679)
Gender		−0.0768 (0.381)		−0.0820 (0.347)	−0.0695 (0.429)
Education		−0.00852 (0.888)		−0.0212 (0.726)	−0.0177 (0.771)
NGO		0.000512 (0.995)		0.00357 (0.967)	−0.00885 (0.919)
Order		0.00660 (0.938)		0.00882 (0.917)	0.00982 (0.908)
ChoiceHU			0.214* (0.023)	0.210* (0.026)	0.213* (0.024)
DecisionRead					−0.0428 (0.739)
_cons	0.236** (0.001)	0.225 (0.313)	−0.0408 (0.610)	−0.0444 (0.841)	0.138 (0.597)

p-values in parentheses
* $= p \leq 0.05$, ** $= p \leq 0.01$, *** $= p \leq 0.001$

Not surprisingly there is also a significant impact on the final decision of the choice of determinant (p-$value = 0.024$) i.e., honesty versus urgency which captures the preferences. In other words, those who claimed to prefer Honesty chose Elephants significantly more than those who preferred Urgency regardless of the condition in which they have been assigned(p-$value = .025$). Everything else being constant, the predicted probability of choosing Elephants is 8.44% (*marginal effect*) higher for an individual who preferred Honesty. All of these effects remained significant when the other variables were included or removed from the regression. None of the control variables (order of the descriptions included) seemed to significantly influence the final decision, which means that the difference in decision-making were essentially not due to the sample hetero-geneity. Similarly, the decision to reread descriptions did not significantly impact the final choice (*p-value* = .740). Both the compatible and incompatible condi-tion significantly lead to a tendency to read the descriptions again (*p-value* = .001 and *p-value* = .006, respectively). In other words, the more information, the more one reads previous information again. We are currently working on further investigation of the data in a companion paper.

4.3 Interpretation

In line with our hypotheses, our results show with no ambiguity that incompat-ible information - that is "distraction" - has a significant impact on the final choice by inducing some extent of switch as compared to both the control group and the compatible information group.

These results are fully consistent with the predictions of the quantum per-suasion model and contradict the predictions of the Bayesian model with respect to the impact of incompatible information. Moreover the fact that general com-patible information had no impact also supports the view that it is not merely "information" that affects the choice because the person is slightly "upset". Instead it is when information induces a change in perspective that something happens even though nothing of relevance is learned.

In addition, the participants' age, gender, level of education or experience with NGOs had no effect on the decision to vote for ECF or TF. The final choice seemed to depend only on the descriptions, the conditions and participants' own beliefs and preferences. We can therefore conclude that our distraction effect – or change of focus – is quite stable among individuals. This supports the hypothesis that the quantum-like structure is a general regularity of the human mind.

The importance of elicited preferences i.e., the answer to "what is determi-nant to your choice" to the final choice underlines that the initial texts were well-understood. The description of the Elephant project was designed to sug-gest more trust to the NGO, while the Tiger project aimed at suggesting higher level of urgency. That explains why respondents who declared Honesty (resp. Urgency) to be determinant were significantly more likely to support the Ele-phant Crisis Fund (resp. Tiger Forever).

The average time to respond to the questionnaire was between 1 and 2 min, which is rather quick. In addition only a tiny proportion of participants (15%) actually used that opportunity to reassess their understanding of the project by rereading the projects descriptions. These two facts support the idea of an absence of conscious reasoning, that is, the respondents did not take time to reflect and reacted spontaneously to the distraction. This is particularly inter-esting for us since the quantum working of the mind is not rational reasoning:

no new information of relevance for the choice was provided and yet it did affect the choice.

5 Discussion and Concluding Remarks

In the experiment we performed, incompatible information i.e., a change of focus, was shown to affect revealed preferences for uncertain alternatives which we interpret as distraction affecting beliefs (rather than preferences). Because the two projects are classical objects our results are not consistent with rationality. When Receiver processes information about a classical object as if it was a quantum system, she is mistaken. But as amply evidenced by Kahneman's best selling book "Thinking Fast and Slow" [11], information processing is not always disciplined by rational thinking when the brain processes information quickly. Moreover a learning process adapted to the quantum-like world may be appropriate when you are interested in actions/decisions produced by other people as they would also have a quantum-like representation of the world. So, while fast quantum-like information processing is inappropriate when dealing with simple decision involving classical objects, it may be suitable in many situations involving human beings. We believe we should not dismiss quantum-like information processing as overly irrational.

Finally, we do not believe that the quantum approach is an alternative to all other behavioral theories. Instead we believe that it provides rigorous foundations to a number of them as argued for instance in [15].

Acknowledgements. We would like to thank Jerome Busemeyer for a very valuable suggestion on the design of the experiment.

References

1. Akerlof, G.A., Shiller, R.J.: Phishing for Phools: The Economics of Manipulation and Deception. Princeton University Press, Princeton (2015)
2. Baron, R.S., Baron, P.H., Miller, N.: The relation between distraction and persuasion. Psychol. Bull. **80**(4), 310–323 (1973)
3. Busemeyer, J.R., Bruza, P.D.: Quantum Models of Cognition and Decision. Cambridge University Press, Cambridge (2012)
4. Chong, D., Druckman, J.N.: Framing theory. Annu. Rev. Polit. Sci. **10**, 103–126 (2007)
5. Danilov, V., Lambert-Mogiliansky, A.: Preparing a (quantum) belief system. Theor. Comput. Sci. **752**, 97–103 (2018)
6. Danilov, V., Lambert-Mogiliansky, A.: Targeting in quantum persuasion problem. J. Math. Econ. **78**, 142–149 (2018)
7. Danilov, V., Lambert-Mogiliansky, A., Vergopoulos, V.: Dynamic consistency of expected utility under non-classical (quantum) uncertainty. Theor. Decis. **84**(4), 645–670 (2018)
8. DellaVigna, S., List, J.A., Malmendier, U.: Testing for altruism and social pressure in charitable giving. Q. J. Econ. **127**(1), 1–56 (2012)

9. Festinger, L., Maccoby, N.: On resistance to persuasive communications. J. Abnorm. Soc. Psychol. **68**(4), 359–366 (1964)
10. Haven, E., Khrennikov, A.: A brief introduction to quantum formalism. In: The Palgrave Handbook of Quantum Models in Social Science, pp. 1–17 (2017)
11. Kahneman, D.: Thinking, Fast and Slow. Macmillan, London (2011)
12. Kamenica, E., Gentzkow, M.: Bayesian persuasion. Am. Econ. Rev. **101**(6), 2590–2615 (2011)
13. Kees, J., Berry, C., Burton, S., Sheehan, K.: An analysis of data quality: professional panels, student subject pools, and Amazon's Mechanical Turk. J. Advert. **46**(1), 141–155 (2017)
14. Kupor, D.M., Tormala, Z.L.: Persuasion, interrupted: the effect of momentary interruptions on message processing and persuasion. J. Consum. Res. **42**(2), 300–315 (2015)
15. Lambert-Mogiliansky, A., Busemeyer, J.: Quantum type indeterminacy in dynamic decision-making: self-control through identity management. Games **3**(2), 97–118 (2012)
16. Petty, R.E., Cacioppo, J.T.: The elaboration likelihood model of persuasion. In: Petty, R.E., Cacioppo, J.T. (eds.) Communication and Persuasion: Central and Peripheral Routes to Attitude Change, pp. 1–24. Springer, New York (1986). https://doi.org/10.1007/978-1-4612-4964-1_1
17. White, L.C., Pothos, E.M., Busemeyer, J.R.: Insights from quantum cognitive models for organizational decision making. J. Appl. Res. Mem. Cogn. **4**(3), 229–238 (2015)

Are Decisions of Image Trustworthiness Contextual? A Pilot Study

Peter D. Bruza[✉] and Lauren Fell

School of Information Systems,
Queensland University of Technology, Brisbane, Australia
p.bruza@qut.edu.au

Abstract. This article documents an empirical pilot study conducted to determine whether decisions of image trustworthiness are contextual. Contextuality is an active area of investigation in quantum cognition, however there has been little compelling evidence of its presence in human information processing. A Bell scenario experimental design was employed which manipulated both content and representational features in order to minimize the difference in marginal probabilities across experimental conditions. In addition, participants were subjected to time pressure in order to promote more spontaneous decisions. Results revealed no significant differences in marginal probabilities, however, no evidence of contextuality was found. The study revealed a tension between the requirement for minimizing the difference in marginal probabilities and the need to produce the strong correlations required to empirically ascertain contextuality.

1 Introduction

Understanding of trust is pivotal in today's environment characterized by claims of fake news and deliberate digital misdirection [15]. It is from concerns regarding the edge of what is acceptable or not acceptable that reputed organizations with image archives such Getty Images (www.gettimages.com) and Reuters (http://www.reuters.com/) have a zero tolerance policy on photo manipulations. A photographer who modified a golfing image to remove a background bystander was terminated by Getty Images in accordance with this policy [12]. Similarly, a Pulitzer prize-winning photographer was fired after he had admitted to altering an image of the conflict in Syria by photoshopping a camera out of the image. In both cases the stance taken is that an image must be a totally true and accurate depiction of reality. Naturally, much hinges on how accuracy is interpreted and where a human subject sets the threshold for an image being "accurate enough" to be judged trustworthy. For example, a human subject might still judge the Pulitzer prize winner's photograph as trustworthy, knowing the camera had been photoshopped out, simply because the object erased did not influence the photograph's resemblance to an actual war scene in Syria. This example attempts to demonstrate that judgments of image trustworthiness are cognitively situated. It turns out that visual fluency is an important factor.

B. Coecke and A. Lambert-Mogiliansky (Eds.): QI 2018, LNCS 11690, pp. 39–50, 2019.
https://doi.org/10.1007/978-3-030-35895-2_3

Visual fluency is based on the principle that any visual stimulus requires cognitive resources to process; the more work required, the less fluent the process. Cognitive work is determined by perceptual processing of image clarity, contrast, etc. [17], and is also determined by the evaluation of aspects of content, resemblance to what is expected, representation (e.g., geometric and artistic depictions), etc [16]. Images that cohere with background beliefs on any of these factors are more visually fluent than properties that surprise or confuse us. The amount of cognitive work can be measured by the speed and accuracy of visual processing as well as the reported ease or difficulty of visual judgments [8]. As a consequence, manipulated images are less detectable the more they conform to largely unconscious rules of visual fluency. Ease of visual processing results in an illusion of accuracy, perhaps because perceptual fluency elicits a feeling of familiarity, and hence trust. For example, [13] found that people are more likely to say that they 'liked' a person shown in an image, if that image was high in visual fluency. Conversely, raucous interruptions to visual fluency are sometimes deployed in image manipulations in order to generate humor, or shock. Recent experimental findings show evidence that both the subject of the image as well as its representational features (features of the image itself as a representation of the subject) were involved when subjects judge the trustworthiness of images such as that of a smiling Vladimir Putin [7]. Although participants in this study were specifically instructed to judge the trustworthiness of an image itself, a dichotomy appeared between participants who were making a deliberative decision based on the content of the image (e.g. Vladimir Putin), and participants making a decision based on representational features of the image, which could be explained in terms of visual fluency. Due to the uncertainty experienced in evaluations of trustworthiness, as well as the potentially high processing difficulty associated with images that were low in visual fluency, the employment of 'Hot' processing was likely in play in this experiment. This system is described as one half of Dual Process Theory, which posits that humans employ two distinct systems for information processing [9]. System 1, or 'Hot' system, is intuitive, fast, and prone to cognitive biases. System 2, or 'Cold' system, is deliberative, slow, and requires significant cognitive resources. The Hot system is often relied upon in circumstances where cognitive resources are sparse [1], uncertainty is present [10], and where a time pressure exists [11].

A conventional way to model the preceding situation is to develop a fusion model whereby an assessment of trust in the content of an image is fused with an assessment of trust in the image as a representation of the content into a single overall assessment of trust. However, is such a reduction into disparate content and representational subsystems valid? This article aims to address this question by determining whether decision making is contextual.

2   A Probabilistic Fusion Model of Trust

Analysis of qualitative feedback in relation to an image of Vladimir Putin[1] revealed a surprising number participants including comments that they didn't trust the image simply because they didn't like Putin, or didn't think he was honest [7]. This was in spite of the fact that subjects were carefully instructed to judge the trustworthiness of the image itself. In other words, these participants seemed to be confounding a decision of whether they trust the image, with a decision with whether they trust the *content* of the image. Conversely, qualitative feedback from other participants revealed that they the trusted the image because they couldn't detect any evidence of manipulation. Such feedback aligns with the visual fluency hypothesis. A similar dichotomy appeared in relation to an image of a strange looking creature known as a frill shark. Both cases are revealing as they seem to indicate that both content and representation features are involved when participants judge the trustworthiness of images. A question that we will explore below is whether decision making around these components are transacted independently, and how that relates to contextuality.

Deciding whether an image if Putin is trustworthy, obviously involves uncertainty, e.g., if may be photoshopped. It is therefore natural to consider a probabilistic decision model. For example, consider the simple model depicted in Fig. 1. The variable S is a random variable which ranged over a set of image stimuli, such as the Putin image. Bivalent random variables C_1, C_2 relate to features associated with the *content* of the image. For example, C_1 may model the decision whether the subject of the image is honest. Conversely, R_1 and R_2 are bivalent random variables that relate to *representational* aspects. For example, R_1 may model the decision whether the image has been manipulated, or not. Variable R_2 might model the decision whether there was something unexpected in the image. Of course, there may be any number of variables related to decisions involving the content and representation of the image. In addition, the variables could be continuous rather than bivalent. However, for simplicity we will use the four bivalent variables; two content based and two representation based. A further assumption if the model latent variables γ and ρ. The C latent variable models the decision whether the content of the image is trustworthy, which only depends variables related to the content. For example, this would equate to the decision whether Putin *the person* is deemed trustworthy. Conversely, the latent R models the decision whether the *image* is deemed to be a true and accurate depiction of reality. Finally, the variable T corresponds to the decision whether the human subject trusts what they have been shown. Such a decision depends on both decisions whether the content and representation are trusted as modeled by the latent variable.

We term such a model, a probabilistic fusion model as both content and representation components are reconciled in order to produce the final decision of trustworthiness. The model allows for one component to dominate decision

---

[1] https://www.newyorker.com/humor/borowitz-report/putin-announces-historic-g1-summit.

making. For example, if the visual fluency is broken, then $p(T = y|\rho)$ would be much higher than $p(T = y|\gamma)$, reflecting that representational aspects of the image dominate decision making of image trustworthiness.

In addition, the following probabilistic relationships are a consequence of the decision model:

$$p(R_1 = y) = p(R_1 = y, C_1 = y) + p(R_1 = y, C_1 = n)$$
$$= p(R_1 = y, C_2 = y) + p(R_1 = y, C_2 = n)$$
$$p(R_2 = y) = p(R_2 = y, C_1 = y) + p(R_2 = y, C_1 = n)$$
$$= p(R_2 = y, C_2 = y) + p(R_2 = y, C_2 = n)$$

and the converse

$$p(C_1 = y) = p(C_1 = y, R_1 = y) + p(C_1 = y, R_1 = n)$$
$$= p(C_1 = y, R_2 = y) + p(C_1 = y, R_2 = n)$$
$$p(C_2 = y) = p(C_2 = y, R_1 = y) + p(C_2 = y, R_1 = n)$$
$$= p(C_2 = y, R_2 = y) + p(C_2 = y, R_2 = n)$$

The preceding probabilistic relationships express that decision making around content and representation do not influence each other. For example, the probability of a decision that a subject trusts Putin (the person), denoted $p(C_1 = y)$, does not vary according to whether subject decides that image has been manipulated, say, denoted $p(C_1 = y, R_1 = y) + p(C_1 = y, R_1 = y)$ or whether they detect something unexpected in the image, denoted $p(C_1 = y, R_2 = y) + p(C_1 = y, R_2 = y)$.

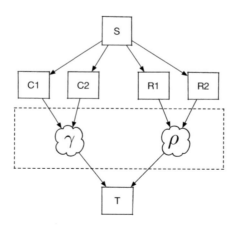

Fig. 1. Probabilistic fusion model of trust

3 Contextuality

A Bell scenario experiment involves two systems C (content) and R (representation). The content system C is probed with two questions modeled by bivalent variables C_1 and C_2 both of which range over the outcomes $\{y, n\}$. Similarly for system R with variables R_1 and R_2. Four measurement contexts are defined by jointly measuring one variable from each system:

$$
C \quad
\begin{array}{cc}
 & \begin{array}{cc} R_1 & R_2 \end{array} \\[2pt]
 & \begin{array}{cccc} y\; n & \quad y\; n \end{array} \\[2pt]
\begin{array}{cc} C_1 & \begin{array}{c} y \\ n \end{array} \end{array} &
\left(\begin{array}{cc|cc}
p_1 & p_2 & p_5 & p_6 \\
p_3 & p_4 & p_7 & p_8 \\
\hline
p_9 & p_{10} & p_{13} & p_{14} \\
p_{11} & p_{12} & p_{15} & p_{16}
\end{array}\right)
\end{array}
\tag{1}
$$

According to the first principle of Contextuality-by-Default, random variables should be indexed according to the experimental conditions in which they are measured [5]. For example, variable C_1 is jointly measured with R_1 in one experimental condition as well as being jointly measured with a variable R_2 in another experimental condition. For this reason, two variables C_{11} and C_{12} are introduced. The same holds for the other three random variables resulting in eight random variables. Their expectations are computed as follows [4]:

$$\langle C_{11} \rangle = 2(p_1 + p_2) - 1 \tag{2}$$
$$\langle C_{12} \rangle = 2(p_5 + p_6) - 1 \tag{3}$$
$$\langle C_{21} \rangle = 2(p_9 + p_{10}) - 1 \tag{4}$$
$$\langle C_{22} \rangle = 2(p_{13} + p_{14}) - 1 \tag{5}$$
$$\langle R_{11} \rangle = 2(p_1 + p_3) - 1 \tag{6}$$
$$\langle R_{12} \rangle = 2(p_9 + p_{11}) - 1 \tag{7}$$
$$\langle R_{21} \rangle = 2(p_5 + p_7) - 1 \tag{8}$$
$$\langle R_{22} \rangle = 2(p_{13} + p_{15}) - 1 \tag{9}$$

Analysis of contextuality in Bell scenario experiments relies on the no-signalling condition. The experience so far in quantum cognition is that it is challenging to design experiments where this condition holds [6]. Moreover, the question that this challenge poses is whether any meaningful conception of contextuality exists when signalling is present. [4] have presented theory that specifies a threshold for signalling below which meaningful contextuality analysis can be performed. Using their approach, the degree of signalling Δ_0 between the content and representation systems is computed as follows:

$$
\Delta_0 = \frac{1}{2}(|\langle C_{11} \rangle - \langle C_{12} \rangle| + |\langle C_{21} \rangle - \langle C_{22} \rangle| + |\langle R_{11} \rangle - \langle R_{12} \rangle| + |\langle R_{21} \rangle - \langle R_{22} \rangle|)
\tag{10}
$$

This sum is essentially based on the sum of differences between marginal probabilities. If $\Delta_0 \geq 1$, then the differences between the marginal probabilities is deemed too great for any meaningful conception of contextuality. When $0 \leq \Delta_0 < 1$, then failure of any of the four inequalities signifies the presence of contextuality. Note that these inequalities revert to the standard CHSH inequalities when signalling is not present ($\Delta_0 = 0$).

$$|\langle C_{11}R_{11}\rangle + \langle C_{12}R_{12}\rangle + \langle C_{21}R_{21}\rangle - \langle C_{22}R_{22}\rangle| \leq 2(1 + \Delta_0) \tag{11}$$
$$|\langle C_{11}R_{11}\rangle + \langle C_{12}R_{12}\rangle - \langle C_{21}R_{21}\rangle + \langle C_{22}R_{22}\rangle| \leq 2(1 + \Delta_0) \tag{12}$$
$$|\langle C_{11}R_{11}\rangle - \langle C_{12}R_{12}\rangle + \langle C_{21}R_{21}\rangle + \langle C_{22}R_{22}\rangle| \leq 2(1 + \Delta_0) \tag{13}$$
$$|-\langle C_{11}R_{11}\rangle + \langle C_{12}R_{12}\rangle - \langle C_{21}R_{21}\rangle + \langle C_{22}R_{22}\rangle| \leq 2(1 + \Delta_0) \tag{14}$$

where

$$\langle C_{11}R_{11}\rangle = (p_1 + p_4) - (p_2 + p_3) \tag{15}$$
$$\langle C_{12}R_{12}\rangle = (p_5 + p_8) - (p_6 + p_7) \tag{16}$$
$$\langle C_{21}R_{21}\rangle = (p_9 + p_{12}) - (p_{10} + p_{11}) \tag{17}$$
$$\langle C_{22}R_{22}\rangle = (p_{13} + p_{16}) - (p_{14} + p_{15}) \tag{18}$$

4 Experiment

4.1 Subjects

Participants consisted of 34 workers using the online crowdsourcing platform, Amazon Mechanical Turk (AMT), and 85 workers using the online crowdsourcing platform, Prolific. Both platforms are used by researchers, amongst others, to post experiments and surveys to a wide audience of thousands of potential participants to view and complete. Participants involved in the present study were able to view any project before agreeing to participate, and were paid a small but standard amount ($1.50 for AMT participants, £1.00 for Prolific participants) per HIT. Both AMT and Prolific platforms allow workers of certain skill, ability or reputation to be specified for a HIT. In this experiment, workers of at least 96% or greater approval rating were selected to balance worker quality with the need to attract a sufficient number of participants. De-identified worker IDs were used to ensure participants did not participate in more than one condition. Participants were sourced from a wide variety of countries, including the US, UK, France, Italy, Australia, New Zealand, and South Africa. A test question was first presented to participants to ensure they could properly view the experiment on their browsers. Data collected from participants who could not correctly answer this test question, did not respond to written segments of the experiment in English, or who were deemed not to have understood instructions based on the relevancy/quality of their responses, were excluded from the study. In total, 3 workers were excluded.

4.2   Design and Materials

Images were selected to fall into two subject categories: untrustworthy subjects and neutral subjects. The first two images depicted infamously untrustworthy subjects: an image of Vladimir Putin[2], and an image of a shark[3]. Figure 2 depicts the images of the neutral subjects: an unknown woman[4], and an antelope with her calf [5]. All images were manipulated in a detectable way. Image presentation was counter-balanced, and each image was accompanied by three responses: A question-pair, a confidence rating, and a reasoning explanation. The question pairs consisted of combinations of the following four questions:

Fig. 2. Image stimuli

- Do you think the person in this image is honest? (for a human image subject)/ Do you think the animal in this image is safe? (for an animal image subject) (Honest/Safe)
- Do think this image has been altered? (Altered)
- Do you think the person in this image is dishonest? (for a human image subject)/ Do you think the animal in this image is unsafe? (for an animal image subject) (Dishonest/Unsafe)
- Do think this image has been unaltered? (Unaltered)

The questions were intended to capture a decision about the subject of the image (human or animal) as well as the features of the image itself (the representation of the subject). The first two questions are the opposite counterparts of the second two. Responses to opposing questions were expected to be unique to each other, based on research showing asymmetry between positive and negatively worded questions [14]. The pairs were combined in the following four ways to form the 4 measurement contexts in a Bell scenario design where two variables (H and D) are related to the content of the image and two variables (A and U) are related to its representation.

---

[2] https://www.iol.co.za/news/world/us-releases-putin-list-13013292.
[3] https://www.complex.com/life/2017/11/scientists-discover-ancient-shark-with-300-teeth-and-people-want-no-part-of-it.
[4] Sourced from Pixabay.
[5] Sourced from Flickr (Igor Shpilenok) under the following license: https://creativecommons.org/licenses/by/2.0/.

- Honest/Safe (H) and Altered (A)
- Honest/Safe (H) and Unaltered (U)
- Dishonest/Unsafe (D) and Altered (A)
- Dishonest/Unsafe (D) and Unaltered (U)

Each image was displayed for 1 s. Placed in a Dual Process context, decisions of trust are assumed to be based on affective characteristics (the 'Hot' system) as well as more rational characteristics (the 'Cold' system). The time pressure was designed to access the 'Hot' system.

4.3   Procedure

Participants were instructed to peruse the task (one condition of the experiment) before agreeing to participate. They were given a test question to ensure they understood instructions and could properly view the images displayed in the task before continuing. This was done by presenting an image of a cat, along with a question asking what animal featured in the image. If the participant could correctly answer this, they were asked to continue to the rest of the task.

Following this question were instructions aimed at introducing the question pair as well as preparing users for the shortness of the image display time. Once understood, users were asked to scroll to the first image. Each image and question group was structured in the following way:

1. A short repeat of image reveal instructions
2. A blank image with the text 'Click to show image' Once clicked, this changed to one of the images outlined earlier, with a number beneath counting down from 1 s. Once 1 s had elapsed, the image disappeared.
3. First response: The full question pair is displayed again (the same question pair as in the instructions and accompanying each image of the HIT), as well as a short version (e.g., Safe/Altered). Beneath this was a multiple choice format with the answers:
   (a) Yes, Yes
   (b) Yes, No
   (c) No, Yes
   (d) No, No
   These were aligned to the short version of the questions to avoid confusion about what each answer referred to.
4. Second response: Participants were then asked to rate, on a slider with a scale from 1 to 7, their confidence in the answers they gave.
5. Third response: Lastly, participants were asked to provide reasoning for the answers they gave in written form. This was to gather further insight into the first response, as well as to determine whether participants correctly understood the questions being asked.

This was repeated for all 4 images, and was followed by a question asking for the country the participants resided in, and a 'Submit' button.

		A		U	
		y	n	y	n
H	y	0.065	0.097	0.194	0.000
	n	0.161	0.677	0.581	0.226
D	y	0.233	0.467	0.516	0.323
	n	0.033	0.233	0.065	0.097

		A		U	
		y	n	y	n
H	y	0.032	0.032	0.097	0.097
	n	0.452	0.484	0.387	0.419
D	y	0.400	0.300	0.323	0.452
	n	0.167	0.133	0.129	0.097

		y	n	y	n
H	y	0.387	0.323	0.355	0.323
	n	0.194	0.097	0.161	0.161
D	y	0.133	0.100	0.129	0.129
	n	0.367	0.400	0.290	0.452

		y	n	y	n
H	y	0.258	0.387	0.581	0.097
	n	0.129	0.226	0.194	0.129
D	y	0.000	0.367	0.323	0.065
	n	0.333	0.300	0.323	0.290

Fig. 3. Pairwise probability distributions corresponding to Eq. (1) for Putin (top left), Shark (top right), Girl (bottom left), antelope (bottom right). Variables H - honest/safe, D - dishonest/unsafe, A -altered, U - unaltered

Table 1. Differences in marginal probabilities: Putin (L) and Shark (R)

		$|\text{diff}|$	$z-\text{score }(p-\text{val})$
0.161	0.194	0.032	-0.894 (0.896)
0.733	0.839	0.105	-0.894 (0.381)
0.226	0.300	0.074	-0.333 (0.751)
0.774	0.581	0.194	1.344 (0.192)

		$|\text{diff}|$	$z-\text{score }(p-\text{val})$
0.065	0.194	0.129	-0.430 (0.742)
0.700	0.774	0.074	-0.566 (0.578)
0.484	0.567	0.083	-0.468 (0.647)
0.484	0.452	0.032	0.174 (0.864)

Table 2. Differences in marginal probabilities: Girl (L) and Antelope (R)

		$|\text{diff}|$	$z-\text{score }(p-\text{value})$
0.710	0.677	0.032	0.229 (0.821)
0.233	0.258	0.025	-0.111 (0.915)
0.581	0.5	0.081	-0.463 (0.649)
0.516	0.419	0.097	0.519 (0.611)

		$|\text{diff}|$	$z-\text{score }(p-\text{value})$
0.645	0.677	0.032	-0.218 (0.830)
0.367	0.387	0.020	-0.101 (0.922)
0.387	0.333	0.053	0.261 (0.799)
0.774	0.645	0.129	0.945 (0.355)

4.4   Results

Tables 1 and 2 show that differences in marginal probabilities vary across all four images. Z–scores were computed to determine whether the differences between marginal probabilities were significant. None of these differences where found to be significant. Table 3 reveals that $\Delta_0 < 1$ for all images which means that the system of CHSH inequalities (11)–(14) can be meaningfully applied to determine the presence of contextuality. The resulting analyses revealed no violations as none of the maximum CHSH values exceeded the degree of contextuality required.

4.5   Discussion

The differences in marginal probabilities reveal that there is signalling between the content and representation decision systems. This is despite the fact the experimental protocol was designed to isolate these sub-systems.

Table 3. CHSH values for the four images

Image	max |CHSH value|	Δ_0	Degree of contextuality
Putin	0.871	0.405	2.811
Shark	0.293	0.318	2.637
Girl	0.293	0.234	2.469
Antelope	1.013	0.235	2.471

Perhaps unsurprisingly, these data reveal that it is challenging to design Bell scenario experiments for cognition where the no signalling condition holds. However, the level of signalling was low enough to meaningfully test for contextuality. In this experiment, no contextuality was found.

The 1 min display time for images employed in the design of the present study was aimed at accessing the Hot system of processing, however, qualitative data revealed evidence that this may have been less affective than anticipated. Participants appeared to dedicate significant time to their explanation of the reasons behind their decisions to a level of detail that reflects the Cold system of processing. Given that contextuality was assumed to reside in the Hot, more affective, processing system, a failure to gather answers based purely on this system may have contributed to the small correlations observed.

When conducting Bell scenario designs, the designer must minimize the differences in marginal probabilities in four cases (See Tables 1 and 2). The only way to do this is to choose appropriate stimuli and questions. Despite judicious choices, it often happens that marginal differences are not minimized in all four cases. The experiment above is yet another example of this. Even when the level of signalling is reduced to meaningfully analyze contextuality ($\Delta_0 < 1$), the requirement for violating the CHSH inequalities is increased by the condition $2(1 + \Delta_O)$. Moreover, in order to secure violations of the CHSH inequalities, the design must produce strong pair wise (anti-)correlations[6] in order to demonstrate the contextuality of the cognitive phenomenon. In previous work [2], we found that the requirement for the experimental design to (1) reduce differences in marginal probabilities and (2) produce strong correlations can quite easily run counter to each other. These two requirements present a significant challenge for experimenters to uncover contextuality using the Bell scenario design. Perhaps these challenges explain part of the reason why demonstrating contextuality in cognition has thus far been so elusive, notwithstanding [3].

5   Summary and Conclusions

This article documents an empirical pilot study to determine whether decisions of image trustworthiness are contextual. A Bell scenario experimental design was

---

[6] Ideally three strong positive correlations and one strong anti-correlation, or the converse.

employed which manipulated both content and representational features from a probabilistic information fusion model in order to minimize the difference in marginal probabilities across experimental conditions. In addition, participants were subjected to time pressure in order to promote judgments emanating from the Hot system of processing. Qualitative data from a previous study suggested evidence that decisions of image trustworthiness were based on both content and representational features of an image. This opened the door for a Bell scenario design based on content and representational decision subsystems. It was hypothesized that there would be little or no signalling the respective subsystems. In addition, two out of the four image stimuli were selected with an expectation of creating large correlations between content and representational features. Two further image stimuli were selected expecting the correlations to be less as a contrast.

Experimental results did confirm our signalling hypothesis to a degree; although signalling was present, it was at statistically insignificant levels. Experiments with larger sample sizes are needed to further confirm this hypothesis.

CHSH analysis, which takes signalling into account, was performed, and no evidence of contextuality was revealed. The expected strong correlations did not eventuate, possibly due to the fact that the Hot processing system, in which contextuality was expected to occur, was not truly accessed. The study revealed a tension between the requirement for minimizing the difference in marginal probabilities and the need to produce the strong correlations required to empirically ascertain contextuality. This experiment highlights the challenges in applying Bell scenario designs for investigating contextuality in cognition.

Acknowledgements. This research was supported by the Asian Office of Aerospace Research and Development (AOARD) grant: FA2386-17-1-4016 and the InterPARES Trust (https://interparestrust.org). Thanks to Abdul Obeid for his technical support.

References

1. Barrett, L.F., Tugade, M.M., Engle, R.W.: Individual differences in working memory capacity and dual-process theories of the mind. Psychol. Bull. **4**(130), 553 (2004)
2. Bruza, P., Kitto, K., Ramm, B., Sitbon, L.: A probabilistic framework for analysing the compositionality of conceptual combinations. J. Math. Psychol. **67**, 26–38 (2015)
3. Cervantes, V., Dzhafarov, E.: Snow queen is evil and beautiful: experimental evidence for probabilistic contextuality in human choices. arXiv:1711.00418v2
4. Dzhafarov, E., Kujala, J.: Probabilistic contextuality in EPR/Bohm-type systems with signaling allowed. In: Dzhafarov, E. (ed.) Contextuality from Quantum Physics to Psychology, pp. 287–308. World Scientific Press, Singapore (2015)
5. Dzhafarov, E.N., Kujala, J.V.: Random variables recorded under mutually exclusive conditions: contextuality-by-default. In: Liljenström, H. (ed.) Advances in Cognitive Neurodynamics (IV). ACN, pp. 405–409. Springer, Dordrecht (2015). https://doi.org/10.1007/978-94-017-9548-7_57

6. Dzhafarov, E., Zhang, R., Kujala, J.: Is there contextuality in behavioral and social systems? Philos. Trans. R. Soc. A **374**(2058), 20150099 (2015)
7. Fell, L., Bruza, P.D., Devitt, K., Oliver, G., Gladwell, M., Partridge, H.: The cognitive decision space of trust: an exploratory study of image trustworthiness and the propensity to deceive. http://eprints.qut.edu.au/102009/ (2016)
8. Jacoby, L., Kelley, C., Dywan, J.: Memory attributions. In: Roediger, H., Craik, F. (eds.) Varieties of Memory and Consciousness: Essays in Honour of Endel Tulving, pp. 391–422. Erlbaum, New York (1989)
9. Kahneman, D.: A perspective on judgment and choice: mapping bounded rationality. Am. Psychol. **9**(58), 697 (2003)
10. Kobus, D.A., Proctor, S., Holste, S.: Effects of experience and uncertainty during dynamic decision making. Int. J. Ind. Ergon. **28**(5), 275–290 (2001)
11. Lipshitz, R., Klein, G., Orasanu, J., Salas, E.: Taking stock of naturalistic decision making. J. Behav. Decis. Making **14**(5), 331–352 (2001)
12. Lum, J.: Getty photographer terminated over altered golf photo. http://petapixel. com/2010/07/19/getty-photographer-terminated-over-altered-golf-photo/
13. Reber, R., Winkielman, P., Schwarz, N.: Effects of perceptual fluency on affective judgments. Psychol. Sci. **9**(1), 45–48 (1998)
14. Rugg, D.: Experiments in wording questions: II. Public Opinion Q. **1**(5), 91 (1941)
15. Shane, S., Goel, V.: Fake Russian facebook accounts bought $100,000 in political ads. https://www.nytimes.com/2017/09/06/technology/facebook-russian-political-ads.html
16. Whittlesea, B.W.: Illusions of familiarity. J. Exp. Psychol. Learn. Mem. Cogn. **19**(6), 1235 (1993)
17. Whittlesea, B.W., Jacoby, L., Girard, K.: Illusions of immediate memory: evidence of an attributional basis for feelings of familiarity and perceptual quality. J. Mem. Lang. **29**(6), 716–732 (1990)

Probabilistic Programs for Investigating Contextuality in Human Information Processing

Peter D. Bruza[1(✉)] and Peter Wittek[2,3,4]

[1] School of Information Systems, Queensland University of Technology,
Brisbane, Australia
p.bruza@qut.edu.au
[2] Rotman School of Management, University of Toronto, Toronto, Canada
peter@peterwittek.com
[3] Creative Destruction Lab, Toronto, Canada
[4] Vector Institute for Artificial Intelligence, Toronto, Canada

Abstract. This article presents a framework for analysing contextuality in human information processing. In the quantum cognition community there has been ongoing speculation that quantum-like contextuality may be present in human cognition. The framework aims to provide a convenient means of designing experiments and performing contextuality analysis in order to ascertain whether this speculation holds. Experimental designs are expressed as probabilistic programs. The semantics of a program are composed from hypergraphs called contextuality scenarios, which, in turn, are used to determine whether the cognitive phenomenon being studied is contextual. Examples are provided illustrate the framework as well as some reflection about its broader application to quantum physics.

1 Introduction

Imagine that you were interested in developing a model of how memory affects the way humans process information. To this end, you design an experiment in which human subjects study a series of words. After the study phase, a cue word is presented. Each subject then recalls the first word that comes to mind from the list just studied. The outcome is recorded as a measurement. How might you go about defining the model? A first step could be to assume that the probability of recall of a word is proportional to its level of activation in memory. So you define a random variable A with this particular function in mind. Furthermore, you adopt a common modelling convention by assuming that the variable A is *independent* of whether it is measured in a laboratory setting or a night club. Contexuality happens when this independence assumption fails.

Contextuality was originally discovered in quantum physics and is heavily studied in quantum information science, where it is closely related to the study of non-locality. It is a subtle issue which impacts the fundamental concept of a

© Springer Nature Switzerland AG 2019
B. Coecke and A. Lambert-Mogiliansky (Eds.): QI 2018, LNCS 11690, pp. 51–62, 2019.
https://doi.org/10.1007/978-3-030-35895-2_4

random variable, which is the basic building block of probabilistic models. This subtlety hinges on whether random variables can justifiably be considered to be independent of the measurement context in which they are used [7]. Within the quantum cognition community, there has been mounting speculation that human information processing is quantum-like contextual, e.g., in human conceptual processing [3,9,18] and perception [4,5,23].

Investigations into contextuality in human cognition have thus far relied on translating experimental designs from quantum physics to human experimental settings, with Bell scenario designs being most often employed. This translation involves quite a tricky process of navigation. Bell scenario designs require the no-signalling condition to hold, however, human cognition, however, seems to be replete with signalling. Fortunately, theoretical results allow contextuality to be meaningfully analysed when signalling is present [12] and successfully applied in [10]. Quantifying the amount of signalling is critical in verifying non-locality in quantum physics and an experimentally relevant formalism has been devised [11], and it is related to the proposal in [12].

A measurement context in quantum physics comprises observables that are jointly measurable, so the order in which the observables within a given context are measured will not affect the associated statistics. This requirement, however, presents an additional challenge for contextuality experiments in human cognition due to the almost ubiquitous presence of order effects.

In the light of preceding challenges, this article presents a framework for conducting contextuality experiments in human cognition. It has two fundamental components. A probabilistic programming language (PPL) which aims to allow experimenters to easily express contextuality experiments. A key feature in this regard are syntactic scopes which allow random variables to be safely overloaded whilst preserving their functional identity. Secondly, the syntactic components of the program are mapped to theoretical construct known as a contextuality scenario. These can be then composed to form a composite contextuality scenario corresponding to the whole program. If there are no probabilistic models that span this composite contextuality scenario, then the phenomenon modelled by the PPL is deemed contextual.

2   A Framework for Determining Contextuality in Human Information Processing

Figure 1 depicts the framework for contextuality analysis. A cognitive phenomenon P is to be studied to determine whether it is contextual. An experimental design is devised in which P is examined in various experimental conditions called "measurement contexts". A measurement context M_i is designed to study P from a particular experimental perspective. The collective measurement contexts aim to reveal a total picture of P in terms of a global model.

The design of the experiment is conceived by the modeller and specified by a PPL. PPLs unify techniques from conventional programming such as modularity, imperative or functional specification, as well as the representation and use of uncertain knowledge. A variety of PPLs have been proposed (see [17] for references), which have attracted interest from artificial intelligence, programming languages, cognitive science, and the natural language processing communities [15]. However, unlike conventional programming languages, which are written with the intention to be executed, a core purpose of a probabilistic program is to specify a model in the form of a probability distribution. In short, PPLs are high-level and universal languages for expressing and computing with probabilistic models in a wide variety of application settings. To the best of our knowledge, PPLs have not yet been applied to investigate contextuality, though some initial ideas in this direction have been put forward [6,7].

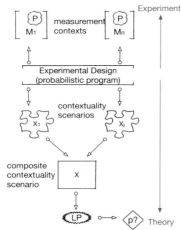

We will now illustrate a probabilistic program (P-program) by means of an example. Order effect experiments involve two measurement contexts each involving two dichotomous variables A and B which represent answers to yes/no questions Q_A and Q_B. In one measurement context, the question Q_A is served before question Q_B and in the second measurement context the reverse order is served, namely Q_B then Q_A. Order effects occur when the answer to the first question influences the answer to the second. These two measurement contexts are syntactically specified by the scopes P1 and P2 shown in Fig. 2.

Fig. 1. Framework for contextuality analysis

```
1   var P1 = context() {
2        var A = flip(0.7)
3        var B = A ? flip(0.8): flip(0.1)
4        var p=[A,B]
5        return {Infer({samples:1000},p}
6   };
7   var P2 = context() {
8        var B = flip(0.4)
9        var A = B ? flip(0.4): flip(0.6)
10       var p=[B,A]
11       return {Infer({samples:1000},p}
12  };
13  return {model(P1,P2)}
```

Fig. 2. Example order effects P-program.

In this P-program, syntax of the form `var B = A ? flip(0.8): flip(0.1)` models the influence of the answer of Q_A on Q_B via a pair of biased coins. In this case, if $Q_A = y$, then the response to Q_B is determined by flipping an 80% biased coin. Conversely, if $Q_A = n$, then the response to Q_B is determined by flipping a 10% biased coin (the choices of such biases are determined by the modeller). It should be noted that the measurement contexts in the order effects program do not reflect the usual understanding of measurement context employed in experiments analyzing contextuality in quantum physics. In these experiments, a measurement context comprises observables that are jointly measurable, so the order in which the observables within a given context are measured will not affect the associated statistics. However, in human information processing experiments order effects are often encountered. Therefore, we take the view that P-progams used to investigate contextuality in human information processing *should* provide syntactic support for measurement contexts where order effects between random variables are occurring.

A number of theoretical frameworks have been developed in order to study contextuality: sheaf theory [1], generalized probability theory [13] and hypergraphs [2]. In the following, hypergraphs are used because their modularity allows a straightforward association with syntactic scopes. The basic theoretical construct is a "contextuality scenario". Definition 2.2.1 [2] states that a *contextuality scenario* is a hypergraph $X = (V, E)$ such that:

- $v \in V$ denotes an outcome which can occur in a measurement context
- $e \in E$ is the set of all possible outcomes given a particular measurement context

The set of hyperedges E are determined by both the measurement contexts as well as the measurement protocol. Each measurement context is represented by an edge in the hypergraph X.

The basic idea is that each syntactic scope in a P-program will lead to a hyperedge, where the events are a complete set of outcomes in the given measurement context specified in the associated scope. Additional hyperedges are a consequence of the constraints inherent in the measurement protocol that is applied. In some cases, hyperedges will have a non-trivial intersection: If $v \in e_1$ and $v \in e_2$, then this represents the idea that the two different measurement outcomes corresponding to v should be thought of as equivalent as will be detailed below by using an order effects experiment as an example.

Contextuality scenarios X_i are composed into a composite contextuality scenario X, which aims to be a total theoretical picture of the phenomenon P. The composition must be sensitive to the experimental design, e.g., the no-signalling condition.

We will now use the order effects example to illustrate the associated contextuality scenario which is shown in Fig. 3.

Firstly, the set of V of events (measurement outcomes) comprises all possible combinations of yes/no answers to the questions Q_A and Q_B, namely $V = \{A = 1 \wedge B = 1, A = 1 \wedge B = 0, A = 0 \wedge B = 1, A = 0 \wedge B = 0\}$, where 1 denotes 'yes' and 0 denotes 'no'.

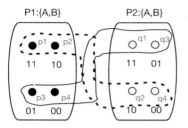

Fig. 3. Contextuality scenario corresponding to P-program depicted in Fig. 2. In total, the hypergraph has 4 edges of four vertices each.

In this figure, the two rounded rectangles represent the events within the two measurement contexts specified by the syntactic scopes P1 and P2. For example, in the rectangle labeled P1, "11" is shorthand for the event $A = 1 \wedge B = 1$, "10" is shorthand for the event $A = 1 \wedge B = 0$, etc. Observe that the corresponding hyperedges (rounded rectangles) contain an exhaustive, mutually exclusive set of events. This is also the case with the two spanning hyperedges going across these rectangles. These spanning edges help illustrate events that are considered to be equivalent.

Firstly, it is reasonable to assume answering yes (or no) to both questions in either measurement context represents equivalent events. Therefore, the events labelled p_1 and p_4 can respectively be assumed equivalent to q_1 and q_4. It becomes a little more subtle when the polarity of the answers differ. For example, the event labelled p_3 represents the event $A = 0 \wedge B = 1$, remembering that question Q_A was asked before question Q_B in this context. The equivalent event in hyperedge P2 is labelled q_2, which corresponds the event $B = 1 \wedge A = 0$, where question B is asked before question A. As conjunction is commutative, it is reasonable to view these two converse events as equivalent. In summary, if p_3 is equivalent to q_2 and p_4 is equivalent to q_4 then the hyperedge $\{p_1, p_2, q_2, q_4\}$ (the dashed hyperedge in Fig. 3) can be established, in addition to the hyperedge $\{p_1, p_2, p_3, p_4\}$.

How to determine contextuality

A *probabilistic model* corresponding to a contextuality scenario X is the mapping of measurement outcomes to a probability $p : V \rightarrow [0, 1]$ (Definition 2.4.1 in [2]). Ref. [19] points out that

> "By defining probabilistic models in this way [rather than by a function $p_e(V)$ depending on the measurement e performed], we are assuming that in the set of experimental protocols that we are interested in, the probability for a given outcome is independent of the measurement that is performed".

Defining probabilistic models in this way formalizes the assumption mentioned in the introduction, namely that random variables are *independent* of the measurement context and thus have a single functional identity. Without a single functional identity it is *impossible* to assign a random variable to represent the outcomes of the same measurement protocol in different measurement contexts.

It is a requirement that the mapping adheres to the expected normalization condition: $\forall_{e \in E} : \sum_{v \in e} p(v) = 1$. By way of illustration, consider once again Fig. 3. This contextuality scenario has four edges. The normalization condition enforces the following constraints:

$$p_1 + p_2 + p_3 + p_4 = 1 \tag{1}$$

$$q_1 + q_2 + q_3 + q_4 = 1 \tag{2}$$

$$p_1 + p_2 + q_3 + q_4 = 1 \tag{3}$$

$$p_3 + p_4 + q_1 + q_2 = 1 \tag{4}$$

where $p_i, 1 \leq i \leq 4$ and $q_j, 1 \leq j \leq 4$ denote the probabilities of outcomes in the four hyperedges. A definition of contextuality can now be presented.

Definition 1 (Probabilistic contextuality). *(General contextuality [2]). Let $X = (V, E)$ be a contextuality scenario. Let $\mathcal{G}(X)$ denote the set of probabilistic models on X. X is deemed "contextual" if $\mathcal{G}(X) = \emptyset$.*

Probabilistic contextuality occurs when there is *no* probabilistic model p corresponding to composite contextuality scenario X. Determining whether X is contextual is computable by a linear program [2].

3 Using Probabilistic Programs to Simulate Bell Scenario Experiments

One of the advantages of using a programming approach to develop probabilistic models is that experimental designs can be syntactically specified in a modular way. In this way, a wide variety of experimental designs across fields can potentially be catered for. For example, consider the situation where an experimenter wishes to determine whether a system S can validly be modelled compositionally in terms of two component subsystems A and B. Two different experiments can be carried out upon each of the two presumed components, which will answer a set of 'questions' with binary outcomes, leading to four measurement contexts. For example, one experimental context would be to ask $A1$ of component A and $B1$ of component B. In Bell scenario experiments, four measurement contexts are typically used: $\{\{A1, B1\}, \{A1, B2\}, \{A2, B1\}, \{A2, B2\}\}$. Bell scenario designs has been widely employed in cognitive psychology to test for contextuality in human cognition [3,9,14,18].

One way to think about system S is that it is equivalent to a set of biased coins A and B, where the bias is local to a given measurement context. Figure 4 depicts a P-program that follows this line of thinking.

```
1   # define the components of the experiment
2   def A = component(A1,A2)
3   def B = component(B1,B2)
4
5   var P1= context(){
6   # declare two binary random variables; 0.5 signifies a fair coin toss
7           var A1 = flip(0.6)
8           var B1 = flip(0.5)
9   # declare joint distribution across the variables A1, B1
10          var p=[A1,B1]
11  # flip the dual coins 1000 times to form the joint distribution
12          return {Infer({samples:1000},p)}
13  };
14  var P2= context(){
15          var A1 = flip(0.4)
16          var B2 = flip(0.7)
17          var p=[A1,B2]
18          return {Infer({samples:1000},p)}
19  };
20  var P3= context(){
21          var A2 = flip(0.2)
22          var B1 = flip(0.7)
23          var p=[A2,B1]
24          return {Infer({samples:1000},p)}
25  };
26  var P4= context(){
27          var A2 = flip(0.4)
28          var B2 = flip(0.5)
29          var p=[A2,B2]
30          return {Infer({samples:1000},p)}
31  };
32  # return a single model
33  return {model({design: 'no-signal',P1,P2,P3,P4})}
```

Fig. 4. Example "Bell scenario" P-program

The Bell scenario program first defines the components A and B together with the associated variables. Thereafter, the program features the four measurement associated contexts P1, P2, P3 and P4. Finally, the line model(design: 'no-signal',P1,P2,P3,P4) specifies that the measurement contexts are to be combined according to the no-signaling condition. The question now to be addressed is how the hypergraph semantics are to be formulated. Reference [2] provides the general semantics of the Bell scenarios by means of multipartite composition of contextuality scenarios.

As these semantics are compositional, it opens the door to map syntactically specified components in a P-program to contextuality scenarios and then to exploit the composition to provide the semantics of the program as a whole.

Consider the Bell scenario program depicted in Fig. 4. The syntactically defined components A and B are modelled as contextuality scenarios X_A and X_B respectively. The corresponding hypergraphs are depicted in Fig. 5.

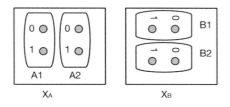

Fig. 5. Contextuality scenarios corresponding to the components A and B defined in the Bell scenario P-program shown in Fig. 4.

Note how the variable definitions associated with the component map to an edge in a hypergraph. For example, the syntax def A = component(A1,A2) corresponds to the two edges labelled $A1$ and $A2$ on the left hand side of Fig. 5.

Contextuality scenarios X_A and X_B are composed into a single contextuality scenario X_{AB}, which will express the semantics of the Bell scenario P-program. However, the no-signalling condition imposes constraints on the allowable probabilistic models on the combined hypergraph structure. Following Definition 3.1.2 in [2], a probabilistic model $p \in \mathcal{G}(X_A \times X_B)$ is a "no signalling" model if:

$$\sum_{w \in e} p(v, w) = \sum_{w \in e'} p(v, w), \forall v \in V(X_A), e, e' \in E(X_B)$$

$$\sum_{w \in e} p(v, w) = \sum_{w \in e'} p(v, w), \forall w \in V(X_B), e, e' \in E(X_A)$$

Reference [2] (p. 45) shows that not all probabilistic models of contextuality scenarios composed by a direct product are "no signalling" models. In order to guarantee that all probabilistic models of a combined contextuality scenario are "no signalling" models, the constituent contextuality scenarios X_A and X_B should be combined by the Foulis-Randall (FR) product denoted $X_{AB} = X_A \otimes^{\mathrm{FR}} X_B$. As with the direct product $X_A \times X_B$ of contextuality scenarios, the vertices of the FR product are defined by $V(X_A \otimes^{\mathrm{FR}} X_B) = V(X_A) \times V(X_B)$. It is with respect to the hyperedges that there is a difference between the FR product and the direct product:

$$X_A \otimes^{\mathrm{FR}} X_B = E_{A \to B} \cup E_{B \leftarrow A}$$

where

$$E_{A \to B} := \bigcup_{v \in e_a} \{v\} \times f(v) : e_a \in E(X_A), f : e_a \to E(X_B)$$

$$E_{A \leftarrow B} := \bigcup_{w \in e_b} f(w) \times \{w\} : e_b \in E(X_B), f : e_b \to E(X_A)$$

We are now in a position to illustrate the semantics of the P-program of Fig. 4 by the corresponding contextuality scenario depicted in Fig. 6. Observe how the FR product produces the extra edges that span the events across measurement contexts labeled P1, P2, P3 and P4. At first these spanning edges may seem arbitrary, but they happen to guarantee that the allowable probabilistic models over the composite contextuality scenario $X_A \otimes^{\mathrm{FR}} X_B$ satisfy the "no signalling" condition [22]. By way of illustration, the normalization condition on edges imposes the following constraints (see Fig. 6):

$$p_1 + p_2 + p_3 + p_4 = 1 \tag{5}$$
$$q_1 + q_2 + q_3 + q_4 = 1 \tag{6}$$
$$p_1 + p_2 + q_3 + q_4 = 1 \tag{7}$$
$$p_3 + p_4 + q_1 + q_2 = 1 \tag{8}$$

where $p_i, 1 \leq i \leq 4$ and $q_j, 1 \leq j \leq 4$ denote the probabilities of events in the respective hyperedges. A consequence of constraints (5) and (7) is that $p_3 + p_4 = q_3 + q_4$. When considering the associated outcomes this means

$$\underbrace{p(A1 = 1 \wedge B1 = 0)}_{p_3} + \underbrace{p(A1 = 1 \wedge B1 = 1)}_{p_4} = \underbrace{p(A1 = 1 \wedge B2 = 0)}_{q_3} + \underbrace{p(A1 = 1 \wedge B2 = 1)}_{q_4}$$

In other words, the marginal probability $p(A1 = 1)$ does not differ across the measurement contexts P1 and P2 specified in the P-program of Fig. 4. In a similar vein, Eqs. (5) and (8) imply that the marginal probability $p(A1 = 0)$ does not differ across measurement contexts P1 and P2. The stability of marginal probability ensures that no signalling is occurring from component B to component A. In quantum physics, the FR product is used to compose contextuality scenarios because this product ensures that there is no signalling between the systems. As a consequence, the operational semantics of the P-program must compute the FR product as some component hyperedges derive from measurement contexts, which have been syntactically specified in the P-program, and other edges express the no-signalling constraint. When the FR product is part of the operational semantics, it provides an underlying data structure which allows both classical and non-classical statistical correlations to be simulated [20]. For example, non-classical correlations between variables such as $\langle A_1 B_1 \rangle$ can be produced by the P-program using standard Bernoulli samplers to produce (biased) coin flips and the underlying hypergraph data structure constrains the sampling to allow quantum-like correlations to emerge.

To illustrate a Bell scenario experiment in human information processing, consider the information fusion model depicted in Fig. 7. The variable S is a random variable which ranges over a set of image stimuli. Human subjects must decide whether an image is trustworthy [8]. Bivalent random variables C_1, C_2 relate to features associated with the *content* of the image. For example, C_1 may model the decision whether a subject deems a person portrayed in an image to be honest. Conversely, R_1 and R_2 are bivalent random variables that relate to *representational* aspects of the image. For example, R_1 may model the decision whether the image has been manipulated, or not. Variable R_2 might model the decision whether there was something unexpected perceived in the image. The latent variable γ models the decision whether the content of the image is trustworthy, and depends on variables related to the content C_1, C_2. Conversely the latent ρ models the decision whether the image is deemed to be authentic, i.e., a true and accurate depiction of reality. Finally, the variable T corresponds to the decision whether the human subject trusts what they have been shown by fusing the assessments regarding the content and representational aspects of the image.

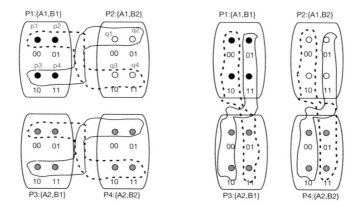

Fig. 6. Contextuality scenario of the P-program of Fig. 4. In total the hypergraph comprises 12 edges of four events. The nodes in rectangles represent events in a probability distribution returned by a given scope: P1, P2, P3, and P4. Note this figure depicts a single hypergraph. Two copies have been made to depict the spanning edges more clearly. This figure corresponds to Figure 7f in [2].

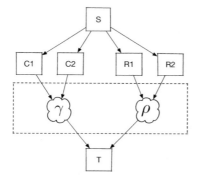

Fig. 7. Probabilistic fusion model of trust

A Bell scenario experiment considers γ and ρ as being separate sub-systems. (See dashed area of Fig. 7). In terms of the framework depicted in 1, four measurement contexts are defined by jointly measuring one variable from each system: $M_1 = \{C_1, R_1\}, M_2 = \{C_2, R_2\}, M_3 = \{C_2, R_1\}, M_4 = \{C_2, R_2\}$.

4   Potential Applications in Quantum Physics

Probabilistic programming languages (PPLs) hav already proved useful in cognitive science [16], but, to our knowledge, they have yet to be seriously taken up by quantum physics. PPLs offer quantum physicists a convenient way to describe specify experiments, and enable a new tool for analyzing statistical correlations based both on simulated as well as actual experimental results. Their potential use is not restricted to Bell scenarios.

Since any PPL is based on random variables, we can ask the question what exactly is a random variable in quantum physics. If we restrict our attention to a single measurement context, due to the normalization constraint, we can think of the measurement context as a (conditional) probability distribution over random variables, which describe the measurement outcomes. More formally, this probability distribution is a normalized measure over the sigma algebra defined by the outcomes. This measure is defined via Born's rule, in other words, the quantum state is embedded in the measure. A Bell scenario is essentially a state preparation protocol where everything that is deterministic is embedded in the quantum state (the unitary operations leading to its preparation), followed by the measurement, which results in stochastic outcomes.

More generally, we can think of a larger system where we only measure a subsystem. This leads to quantum channels, which are described by completely positive trace preserving maps. A quantum channel, however, must be deterministic in the sequence of events, and, for instance, a measurement choice at a later time step cannot depend on the outcome of a previous measurement. We must factor in such classical and quantum memory effects, as well as the potentially indeterminate causal order of events. The quantum comb or process matrix [21] formalism addresses these more generic requirements. Either formalism introduces a generalized Born rule, where deterministic and stochastic parts of the system clearly separate, and thus give a clear way of defining random variables. PPLs offer potential in expressing models designed in these frameworks.

Acknowledgements. This research was supported by the Asian Office of Aerospace Research and Development (AOARD) grant: FA2386-17-1-4016.

References

1. Abramsky, S., Brandenburger, A.: The sheaf-theoretic structure of non-locality and contextuality. New J. Phys. **13**, 113036 (2011)
2. Acin, A., Fritz, T., Leverrier, A., Sainz, A.: A combinatorial apporoach to nonlocality and contextuality. Commun. Math. Phys. **334**, 533–628 (2015)
3. Aerts, D., Gabora, L., Sozzo, S.: Concept combination, entangled measurements, and prototype theory. Top. Cogn. Sci. **6**, 129–137 (2014)
4. Asano, M., Hashimoto, T., Khrennikov, A., Ohya, M., Tanaka, Y.: Violation of contextual generalization of the Leggett–Garg inequality for recognition of ambiguous figures. Physica Scripta (T163), 014006 (2014)
5. Atmanspacher, H., Filk, T.: A proposed test of temporal nonlocality in bistable perception. J. Math. Psychol. **54**, 314–321 (2010)
6. Bruza, P.D.: Syntax and operational semantics of a probabilistic programming language with scopes. J. Math. Psychol. **74**, 46–57 (2016)
7. Bruza, P.D.: Modelling contextuality by probabilistic programs with hypergraph semantics. Theor. Comput. Sci. **752**, 56–70 (2017)
8. Bruza, P., Fell, S.: Are decisions of image trustworthiness contextual? A pilot study. In: Lambert-Mogilliansky, A., Coecke, B. (eds.) Quantum Interaction: 11th International Conference (QI 2018). Lecture Notes in Computer Science. Springer, Heidelberg (2018)
9. Bruza, P., Kitto, K., Ramm, B., Sitbon, L.: A probabilistic framework for analysing the compositionality of conceptual combinations. J. Math. Psychol. **67**, 26–38 (2015)
10. Cervantes, V., Dzhafarov, E.: Snow queen is evil and beautiful: experimental evidence for probabilistic contextuality in human choices. arXiv:1711.00418v2

11. Chaves, R., Kueng, R., Brask, J.B., Gross, D.: Unifying framework for relaxations of the causal assumptions in Bell's theorem. Phys. Rev. Lett. **114**(14), 140403 (2015)
12. Dzhafarov, E., Kujala, J.: Probabilistic contextuality in EPR/Bohm-type systems with signaling allowed. In: Dzhafarov, E. (ed.) Contextuality from Quantum Physics to Psychology, chap. 12, pp. 287–308. World Scientific Press (2015)
13. Dzhafarov, E., Kujala, J., Larsson, J.: Contextuality in three types of quantum-mechanical systems. Found. Phys. **7**, 762–782 (2015)
14. Dzhafarov, E., Zhang, R., Kujala, J.: Is there contextuality in behavioral and social systems? Philos. Trans. Roy. Soc. A **374**, 20150099 (2015)
15. Goodman, N.D., Stuhlmüller, A.: The design and implementation of probabilistic programming languages (2014). http://dippl.org. Accessed 14 Sept 2017
16. Goodman, N.D., Tenenbaum, J.B.: Probabilistic Models of Cognition (2016). http://probmods.org/v2. Accessed 5 June 2017
17. Gordon, A., Henzinger, T., Nori, A., Rajamani, S.: Probabilistic programming. In: Proceedings of the on Future of Software Engineering (FOSE 2014), pp. 167–181 (2014)
18. Gronchi, G., Strambini, E.: Quantum cognition and Bell's inequality: a model for probabilistic judgment bias. J. Math. Psychol. **78**, 65–75 (2016)
19. Henson, J., Sainz, A.: Macroscopic noncontextuality as a principle for almost-quantum correlations. Phyical Rev. A **91**, 042114 (2015)
20. Obeid, A., Bruza, P.D., Wittek, P.: Evaluating probabilistic programming languages for simulating quantum correlations. PLoS One **14**(1), e0208555 (2019)
21. Oreshkov, O., Costa, F., Brukner, C.: Quantum correlations with no causal order. Nat. Commun. **3**, 1092 (2012)
22. Sainz, A., Wolfe, E.: Multipartite composition of contextuality scenarios. arXiv:1701.05171 [quant-ph] (2017)
23. Zhang, R., Dzhafarov, E.N.: Testing contextuality in cyclic psychophysical systems of high ranks. In: de Barros, J.A., Coecke, B., Pothos, E. (eds.) QI 2016. LNCS, vol. 10106, pp. 151–162. Springer, Cham (2017). https://doi.org/10.1007/978-3-319-52289-0_12

Episodic Source Memory over Distribution by Quantum-Like Dynamics – A Model Exploration

J. B. Broekaert(✉) [ID] and J. R. Busemeyer [ID]

Department of Psychological and Brain Sciences, Indiana University, Bloomington, USA
{jbbroeka,jbusemey}@indiana.edu

Abstract. In *source memory* studies, a decision-maker is concerned with identifying the context in which a given episodic experience occurred. A common paradigm for studying source memory is the 'three-list' experimental paradigm, where a subject studies three lists of words and is later asked whether a given word appeared on one or more of the studied lists. Surprisingly, the sum total of the acceptance probabilities generated by asking for the source of a word separately for each list ('list 1?', 'list 2?', 'list 3?') exceeds the acceptance probability generated by asking whether that word occurred on the union of the lists ('list 1 or 2 or 3?'). The episodic memory for a given word therefore appears *over distributed* on the disjoint contexts of the lists. A quantum episodic memory model [QEM] was proposed by Brainerd, Wang and Reyna [8] to explain this type of result. In this paper, we apply a Hamiltonian dynamical extension of QEM for over distribution of source memory. The Hamiltonian operators are simultaneously driven by parameters for re-allocation of *gist*-based and *verbatim*-based acceptance support as subjects are exposed to the cue word in the first temporal stage, and are attenuated for description-dependence by the querying probe in the second temporal stage. Overall, the model predicts well the choice proportions in both separate list and union list queries and the over distribution effect, suggesting that a Hamiltonian dynamics for QEM can provide a good account of the acceptance processes involved in episodic memory tasks.

Keywords: Recognition memory · Over distribution · Quantum modeling · Word list · Verbatim · Gist

1 Familiarity and Recollection, Verbatim and Gist

Recognition memory models predict judgments of 'prior occurrence of an event'. In recognition, Mandler distinguished a *familiarity* process and a *retrieval* - or *recollection* - process that would evolve separately but also additively [21]. The familiarity of a memory would relate to an 'intra event organizational integrative

© Springer Nature Switzerland AG 2019
B. Coecke and A. Lambert-Mogiliansky (Eds.): QI 2018, LNCS 11690, pp. 63–75, 2019.
https://doi.org/10.1007/978-3-030-35895-2_5

process', while retrieval relates to an 'inter event elaborative process'. Extending this *dual process* modeling work, by Tulving [26] and Jacoby [17], a 'conjoint recognition' model was developed by Brainerd, Reyna and Mojardin [4] which provides separate parameters for the entangled processes of identity judgement, similarity judgment and response bias. Their model implements *verbatim* and *gist* dimensions to memories. Verbatim traces hold the detailed contextual features of a past event, while gist traces hold its semantic details. In recognition tasks we would access *verbatim* and *gist* trace in parallel. The verbatim trace of a verbal cue handles it surface content like orthography and phonology for words with its contextual features like in this case, colour of back ground and text font. The verbal cue's gist trace will encode relational content like semantic content for words, also with its contextual features. This development recently received a quantum formalisation for its property of superposed states to cope with *over distribution* in memory tests [8,9,14]. In specifically designed expermental tests it appeared episodic memory of a given word is *over distributed* on the disjoint contexts of the lists, letting the acceptance probability behave as a *subadditive* function [6,9].

Quantum-Like Memory Models. The Quantum Episodic Memory model (QEM) was proposed by Brainerd, Wang and Reyna [8]. It assumes a Hilbert space representation in which verbatim, gist, and non-related components are orthogonal, and in which recognition engages the gist trace in target memories as well. We will provide ample detail about this model in the next section, since our dynamical extension is implemented in essentially the same structural setting. QEM was extended to generalized-QEM (GQEM) by Trueblood and Hemmer to model for *incompatible* features of gist, verbatim and non-related traces [25]. Subjacent is the idea that these features are serially processed, and that gist precedes verbatim since it is processed faster. Independently, Denolf and Lambert-Mogiliansky have considered the accessing of gist and verbatim as incompatible process features. This aspect is implemented in an intrinsically quantum-like manner in their *complementarity*-based model for Complementary Memory Types (CMT) [15,16,20]. We previously developed a Hamiltonian dynamical extension of QEM for *item* memory tasks [11]. The dynamical formalism allows to describe time development of the acceptance decision based on gist, verbatim and non-related traces. Finally, also a semantic network approach by Bruza, Kitto, Nelson, and McEvoy [23] was developed in which the target word is adjacent to its associated terms and the network is in a quantum superposition state of either complete activation or non-activation (see also [12]).

We note that dynamical approaches to quantum-like models have been proposed previously, e.g. in decision theory by Busemeyer and Bruza [14], Pothos and Busemeyer [24], Martínez-Martínez [22], Kvam et al. [19], Busemeyer et al. [13] and by Yearsley and Pothos [28], in cognition by Aerts et al. [2], in perception theory by Atmanspacher and Filk [3]. An overview of quantum modelling techniques is given in Broekaert et al. [10].

2   True Memory, False Memory, over Distributed Memory

In the *conjoint process dissociation* model (CPD) a sufficient parametrisation is present to capture the four distinct response patterns of true, false, over distributed and forgotten memories of the three-list paradigm. The precise identification of the types of memories for a given target requires a composite outcome for the acceptance to three lists at once (see Fig. 1 and Table 1). For instance, should a participant report "the word appeared on L_1, on L_2 but not on L_3," when the cue word came from list L_2 then this participant clearly showed a case of memory over distribution. If however that same answer had been given for a cue coming from list L_3 then this participant showed a case of false memory.

The participant is however well informed at the start that the word lists do not overlap. It makes therefor no sense to ask for an answer to a conjunctive composition query probe at one instance: multiple-yes answers would be absent and therefor no cases of over distribution could be produced. A quantum based model for the *conjunction* of queries moreover requires a procedure specification for its formal representation, since measurement outcomes in quantum models are sensitive to ordering of the measurement operators for non-compatible questions [1, 14, 24, 27]. While the projectors for list membership, Eq. 6, are commutative the dynamical process between two measurements will void that order invariance, as we will see in the next section. The dynamical process implies that the Hamiltonian-QEM predicts different acceptance probabilities for different query orderings, e.g. $p(L_i? \circ L_j?|L_i) \neq p(L_j? \circ L_i?|L_i)$. It is therefor not possible in Hamiltonian-QEM to define a unique expression for expressions like $p(\neg L_i? \cap L_j? \cap L_k?|L_i)$ (cfr [8]) without additional information on the order of querying.

L_i? L_j? L_k? | L_i

yes	yes	yes	→	over distribution
yes	yes	no	→	over distribution
yes	no	yes	→	over distribution
yes	no	no	→	true memory
no	yes	yes	→	false memory
no	yes	no	→	false memory
no	no	yes	→	false memory
no	no	no	→	forgotten

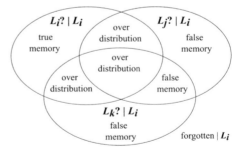

Fig. 1. Logic of false memory, true memory and over distribution in the three list paradigm for source memory, for a target which is a studied word from list L_i. Indices $[i, j, k]$ are permutations of $[1, 2, 3]$. For a distractor, which is an unstudied word from L_4, all response triplets are erroneous memories, except the triple 'no' which is a correct no-memory evaluation.

In the three list experimental paradigm the query probes are kept separated – 'did the word appear on List 1' (L_1?), 'did the word appear on List 2' (L_2?), 'did the word appear on List 3' (L_2?) and the disjunctive probe 'did the word appear on one of the lists' (L_{123}?) – and are randomised between other acceptance tasks for other words [8, 16, 18, 25].

From a classical set theoretic perspective we can relate the acceptance probability for the disjunctive probe with the acceptance probabilities of the single probes;

$$|S(L_1 \cup L_2 \cup L_3?)| = \sum_i |S(L_i?)| - \sum_{i<j} |S(L_i \cap L_j?)| + |S(L_1 \cap L_2 \cap L_3?)| \tag{1}$$

(for a word from a given set L_k). Then we define the unpacking factor UF as the ratio of number of acceptance responses for the separate queries per list over the number of cases of the query for the joined lists:

$$UF(k) = \frac{\sum_i |S(L_i)|}{|S(L_1 \cup L_2 \cup L_3?)|} \tag{2}$$

In terms of summed acceptance probabilities and taking into account classical set relation, Eq. (1), and some algebra, the interpretation of the unpacking factor is apparent;

$$
\begin{aligned}
UF(k) &= \frac{p(L_1?|L_k) + p(L_2?|L_k) + p(L_3?|L_k)}{p(L_{123}?|L_k)} \\
&= 1 + \frac{p(L_1 \cap L_2 \cap \neg L_3?|L_k) + p(\neg L_1 \cap L_2 \cap L_3?|L_k) + p(L_1 \cap \neg L_2 \cap L_3?|L_k)}{p(L_{123}?|L_k)} \\
&\quad + 2\,\frac{p(L_1 \cap L_2 \cap L_3?|L_k)}{p(L_{123}?|L_k)}
\end{aligned}
\tag{3}
$$

For every index value k of the target's list, the excess value of UF above 1 is caused by three over distribution terms of which the 'always accept' term is double weighted, and one false memory term of the type 'accept on all lists except the true source'. For example, when $k = 1$, the term $\frac{p(\neg L_1 \cap L_2 \cap L_3?|L_1)}{p(L_{123}?|L_1)}$ relates to the case a target from L_1 was not accepted on that list while it was accepted both on L_2 and L_3, constituting a false memory contribution to UF.

Since the lists in the experimental design are disjoint, according to classical logic the right hand side is equal to 1. For experimental choice proportions however the unpacking factor shows to be significantly larger than 1 [8]. Modulo the fact that along three terms for over distribution the unpacking factor always mixes in one term of false memory as well, we will still use the unpacking factor as a measure for over distribution, besides its correct measure for subadditivity.[1]

---

[1] In principle contributions of false memories and over distributions could be fully separated if one would measure the acceptance probabilities for disjunctions of all disjunctive list *pairs* as well.

3   The Hamiltonian Based QEM Model

In essence the Hamiltonian based QEM model describes in two subsequent temporal stages how the belief state of the participants evolves through the experimental paradigm. This change of the belief state is described by two distinct Schrödinger evolutions. In the first stage the participant is presented with a cue originating from one of the four lists - three of them with targets, one with distractors. First, the participant processes this incoming information to change her initial 'uniform' belief state into a state informed by the presented cue and her memory. Subsequently this state of belief is then further evolved due to the processing of the information of the probe. The information of the probe allows for a response bias or description-dependence. We expect the latter evolution to be an attenuation of the first stage recognition phase.

State Vectors. In line with Brainerd, Wang and Reyna's development of QEM, the state vector is expressed on the orthogonal basis (V_1, V_2, V_3, G, N). The model thus provides a dedicated dimension for verbatim support for each list, and a dimension for gist support shared for all lists. The last dimension is dedicated to support for non-related items or distractors. In our dynamical development of QEM each state vector is modulated according the cue and probe combination to which the subject is exposed

$$\Psi_{probe|cue}(t) = [\psi_{p|c_{V_1}}(t), \psi_{p|c_{V_2}}(t), \psi_{p|c_{V_3}}(t), \psi_{p|c_G}(t), \psi_{p|c_N}(t)]^\tau,$$

amounting to sixteen distinct states in the present experimental paradigm.

The Initial State Vector. In the basis (V_1, V_2, V_3, G, N), the generalized initial state vector is expressed as

$$\Psi_0(g) = \left[\sqrt{(1-g^2)/6}, \sqrt{(1-g^2)/6}, \sqrt{(1-g^2)/6}, g/\sqrt{2}, 1/\sqrt{2} \right]^\tau \qquad (4)$$

where we restrict the parameter $g \in [-1, 1]$. This initial belief state of the subject reflects to certain extent the fact - of which the subject is informed - that in this experiment half of the cues are non-studied, and half of them come from the three studied lists (p. 419, [7]). This form also implements the idea that the subject at the start has a latent tendency for acceptance of the cues. This form of the initial state expresses that a cue from a studied list elicits on average $(1 - g^2)/6$ acceptance probability from verbatim trace and $g^2/2$ acceptance probability from the gist trace.[2] The parameter g therefor indicates the preponderance of gist in

---

[2] Without taking into account of dynamics for the effect of cue or probe, but still applying the measurement projections Eq. (6), the amount of gist g in the initial state shows a latent tendency for overdistribution

$$p_0(L1) + p_0(L2) + p_0(L3) + p_0(N) = 1 + g^2$$

with $p_0(N) = 1/2$. Clearly the experimental description 'half of the cues are N, the other half originate from the lists', cannot be implemented exactly due to overdistribution.

the initial state for a given cue. It should be noticed however that the initial state itself is not measured on. In the Hamiltonian model a dynamical evolution transforms the initial belief state till the point of measurement. Our present assessment of the initial component amplitudes therefore only concern a latent tendency. We will at present fix the initial state to correspond mathematically to equal weighting of acceptance support by verbatim trace and by gist trace for the three studied lists. The initial state then reads, with $g = 0.5$ in Eq. (5),

$$\Psi_0 = \left[1/2\sqrt{2}, 1/2\sqrt{2}, 1/2\sqrt{2}, 1/2\sqrt{2}, 1/\sqrt{2}\right]^\tau \tag{5}$$

Measurement Projectors. The subject's response to the probes L_i?, or L_{123}? for a given cue of L_j, are obtained by applying the measurement operators on the final state $\Psi_{probe|cue}$. The measurement operators are projector matrices which select the components of the final outcome vectors for the specific response. The projector for e.g. L_1? must select both the dedicated verbatim amplitude $\psi_{L_1|c_{V_1}}$ and the gist component $\psi_{L_1|c_G}$. In (V_1, V_2, V_3, G, N) ordered Hilbert space the corresponding projector M_{L_1}? will thus collapse the outcome state to a subspace spanned on the (V_1, G) basis, and is implemented by a matrix with diagonal elements $(1, 0, 0, 1, 0)$ and 0 otherwise. All projectors for the three list paradigm are implemented accordingly:

$$M_{L_1?}=\begin{bmatrix}1&0&0&0&0\\0&0&0&0&0\\0&0&0&0&0\\0&0&0&1&0\\0&0&0&0&0\end{bmatrix}, M_{L_2?}=\begin{bmatrix}0&0&0&0&0\\0&1&0&0&0\\0&0&0&0&0\\0&0&0&1&0\\0&0&0&0&0\end{bmatrix}, M_{L_3?}=\begin{bmatrix}0&0&0&0&0\\0&0&0&0&0\\0&0&1&0&0\\0&0&0&1&0\\0&0&0&0&0\end{bmatrix}, M_{L_{123}?}=\begin{bmatrix}1&0&0&0&0\\0&1&0&0&0\\0&0&1&0&0\\0&0&0&1&0\\0&0&0&0&0\end{bmatrix} \tag{6}$$

We notice the acceptance projectors commute, but are not orthogonal as they all include the gist component in their projective subspace;

$$\left[P_{L_i?}, P_{L_j?}\right] = 0, \; P_{L_i?}.P_{L_j?} \neq 0 \qquad (i \neq j)$$

Trueblood and Hemmer [25], and Denolf and Lambert-Mogliansky [16] point out QEM's orthogonality of verbatim, gist and non-related features need not necessarily be retained (cfr model specifications in Sect. 1). In our dynamical extension of QEM a query probe engenders its proper dynamics, therefor repeated application of projectors, without intermediate evolution, will not occur.

Acceptance Probabilities. In quantum-like models probabilities are given by the squared length of the projected outcome vectors. Using the projector operators, Eq. (6), the resulting acceptance probabilities for a given probe L_i? after a given cue L_j, are explicitly given by the expressions:

$$p(L_i?|L_j) = \left|\psi_{L_i?|L_j\,V_i}\right|^2 + \left|\psi_{L_i?|L_j\,G}\right|^2, \tag{7}$$

$$p(L_{123}?|L_j) = \left|\psi_{L_{123}?|L_j\,V_1}\right|^2 + \left|\psi_{L_{123}?|L_j\,V_2}\right|^2 + \left|\psi_{L_{123}?|L_j\,V_3}\right|^2 + \left|\psi_{L_{123}?|L_j\,G}\right|^2 \tag{8}$$

Notice that the probe index i runs from 1 to 3, while the cue index j runs from 1 to 4 since it includes the non-studied list L_4.

With the explicit expressions of the acceptance probabilities we obtain, using some algebra, the potential for super-additivity in the Hamiltonian-QEM model by means of the unpacking factor, for a given cue c;

$$UF(c) = 1 + \frac{\left|\psi_{L_1?|c_{V_1}}\right|^2 - \left|\psi_{L_{123}?|c_{V_1}}\right|^2 + \left|\psi_{L_1?|c_{V_2}}\right|^2 - \left|\psi_{L_{123}?|c_{V_2}}\right|^2 + \left|\psi_{L_1?|c_{V_3}}\right|^2 - \left|\psi_{L_{123}?|c_{V_3}}\right|^2}{\left|\psi_{L_{123}?|c_{V_1}}\right|^2 + \left|\psi_{L_{123}?|c_{V_2}}\right|^2 + \left|\psi_{L_{123}?|c_{V_3}}\right|^2 + \left|\psi_{L_{123}?|c_G}\right|^2}$$
$$+ \frac{\left|\psi_{L_1?|c_G}\right|^2 + \left|\psi_{L_2?|c_G}\right|^2 + \left|\psi_{L_3?|c_G}\right|^2 - \left|\psi_{L_{123}?|c_G}\right|^2}{\left|\psi_{L_{123}?|c_{V_1}}\right|^2 + \left|\psi_{L_{123}?|c_{V_2}}\right|^2 + \left|\psi_{L_{123}?|c_{V_3}}\right|^2 + \left|\psi_{L_{123}?|c_G}\right|^2} \tag{9}$$

From which follows that - given the positive gist balance in the second fraction - most often the Hamiltonian-QEM model will predict subadditivity in the three list paradigm. However, like in the Hamiltonian QEM model for the *item false memory* paradigm of Broekaert and Busemeyer [11], we find the model can in principle also account for cases that satisfy the additivity of disjoint sub-events, or violate it in super-additive manner.

Hamiltonians. The dynamical evolution of the state vectors is determined by the Hamiltonian operators. The Hamiltonian reflects the cognitive processing which is engendered by the information in the word cue - differently for a cue from the studied lists or the unrelated list. In the present three-list paradigm these operators are constructed along four transports - or re-allocations - between components of the belief state; (i) between gist-based component and non-cue verbatim-based component ($G \leftrightarrow V_{\neg i}$), (ii) between gist-based component and non-related component ($G \leftrightarrow N$), (iii) between cue verbatim-based component and non-cue verbatim-based component ($V_i \leftrightarrow V_{\neg i}$), and (iv) between cue verbatim-based component and non-related component ($V_i \leftrightarrow N$). The Hamiltonians are constructed by combining the off-diagonal parametrised Hadamard gates of each transport [11]. The Hamiltonian parameter γ controls the transport of acceptance probability amplitude from non-cue verbatim-based components towards the gist-based component, or back. Similarly γ' regulates transport of acceptance probability amplitude of the non-related component from or towards the gist-based component. Precisely the strength γ' of this dynamic will be made use of to adjust for the four distinct types of word cues 'high frequency & concrete' (HFC), 'high frequency & abstract' (HFA), 'low frequency & concrete' (LFC), 'low frequency & abstract' (LFA). Brainerd and Reyna suggest abstract words have weaker verbatim traces than concrete words and low-frequency words have weaker verbatim traces than high-frequency words [5,7,8]. A tendency which suggests gist based transport from the non-related component to vary—namely $\{\gamma'_{HFC}, \gamma'_{HFA}, \gamma'_{LFC}, \gamma'_{LFA}\}$—for these distinct types of word cues.

The parameter ν controls transport of acceptance probability amplitude between non-cue verbatim-based components and the cue verbatim-based component. Similarly ν' controls the transport of the acceptance probability amplitude of the non-related component from or towards the cue verbatim-based component.

In our present development we demonstrate the parameters fulfil their intended transport: the parameters γ and γ' for gist-based transport, and ν and ν' for cue verbatim-based transport. We could have implemented an effect of *list order* by distinguishing for each list the transports of the non-related component $(\gamma'_{i,Abs.type}, \nu'_i)$ for forgetfulness variation per list, or by distinguishing the transports of the target verbatim-based component (γ_i, ν_i) for gist-verbatim diversified acceptance variation per list.

Diminishing the strength of transports, by quenching the driving parameters, will let outcome acceptance beliefs tend to concur with the acceptance beliefs inherent to the initial state.

First Temporal Stage. The presentation of the word cue starts the memory process of recollection and familiarity in the subject. For cues from studied lists the hamiltonians H_{1c}, H_{2c} and H_{3c} engage verbatim-based and gist-based belief. For non-studied cues from list 4, the hamiltonian H_{4c} engages for the non-related belief.

$$H_{1c}(\nu, \nu', \gamma, \gamma') = \begin{bmatrix} 1 & \nu & \nu & 0 & \nu' \\ \nu & -1 & 0 & \gamma & 0 \\ \nu & 0 & -1 & \gamma & 0 \\ 0 & \gamma & \gamma & 1 & \gamma' \\ \nu' & 0 & 0 & \gamma' & -1 \end{bmatrix}, \quad H_{2c}(\nu, \nu', \gamma, \gamma') = \begin{bmatrix} -1 & \nu & 0 & \gamma & 0 \\ \nu & 1 & \nu & 0 & \nu' \\ 0 & \nu & -1 & \gamma & 0 \\ \gamma & 0 & \gamma & 1 & \gamma' \\ 0 & \nu' & 0 & \gamma' & -1 \end{bmatrix}, \quad (10)$$

$$H_{3c}(\nu, \nu', \gamma, \gamma') = \begin{bmatrix} -1 & 0 & \nu & \gamma & 0 \\ 0 & -1 & \nu & \gamma & 0 \\ \nu & \nu & 1 & 0 & \nu' \\ \gamma & \gamma & 0 & 1 & \gamma' \\ 0 & 0 & \nu' & \gamma' & -1 \end{bmatrix}, \quad H_{4c}(\nu', \gamma') = \begin{bmatrix} -1 & 0 & 0 & 0 & \nu' \\ 0 & -1 & 0 & 0 & \nu' \\ 0 & 0 & -1 & 0 & \nu' \\ 0 & 0 & 0 & -1 & \gamma' \\ \nu' & \nu' & \nu' & \gamma' & 1 \end{bmatrix}. \quad (11)$$

Second Temporal Stage. The presentation of the probe starts the subsequent decision process in the subject. The outcome of the cue processing and the subsequent probe can be confirmatory or dissonant. When probing for a studied list, the Hamiltonian H_{ip} for a probe L_i? is equated to the Hamiltonian H_{ic} for processing a cue from L_i, in which the driving parameters are now attenuated by a factor κ. When probing for the union of lists, the dedicated Hamiltonian H_{123p} for probe L_{123}? is equated to the sum of the separate Hamiltonians H_{ip} for processing of studied cues from L_i, in which the driving parameters are again attenuated by the factor κ. This separable construction of the Hamiltonian reflects the cognitive processing of the union list consists of the parallel processing of membership to the three separate lists, and results in the summed attenuated transport between the non-related component and equally all verbatim components and the gist component.

$$H_{1p}(\nu, \nu', \gamma, \gamma', \kappa) = H_{1c}(\kappa\nu, \kappa\nu', \kappa\gamma, \kappa\gamma') \tag{12}$$

$$H_{2p}(\nu, \nu', \gamma, \gamma', \kappa) = H_{2c}(\kappa\nu, \kappa\nu', \kappa\gamma, \kappa\gamma') \tag{13}$$

$$H_{3p}(\nu, \nu', \gamma, \gamma', \kappa) = H_{3c}(\kappa\nu, \kappa\nu', \kappa\gamma, \kappa\gamma') \tag{14}$$

$$H_{123p}(\nu, \nu', \gamma, \gamma', \kappa) = H_{1p} + H_{2p} + H_{3p} \tag{15}$$

Parameters. In recapitulation; the Hamiltonian QEM model uses 5 parameters $\{\mu, \mu', \gamma, \gamma', \kappa\}$ to describe the dynamics of the subject's belief state to evolve from her prior partially informed expectation to the final decision of acceptance of the cue, accounting for 16 acceptance probabilities $p(L_i?|L_j)$. In Sect. (4), we fitted the parameters for the four distinct types of frequency and concreteness of the cue words HFC, HFA, LFC and LFA. Between these sets we only changed the gist transport parameter γ' ($G \leftrightarrow N$) in the Hamiltonians, respectively γ'_{HFA}, γ'_{HFA}, γ'_{LFC} and γ'_{LFA}, and otherwise maintained the same Hamiltonian drivers γ, ν, ν' and κ. In total therefor 8 parameters are needed to predict the 64 values of the acceptance probabilities.

Unitary Time Evolution. In quantum-like models the Hamiltonian is the operator for infinitesimal time change of the belief state. The operator $U(t)$ for time propagation over a time range t is given by

$$U(t) = e^{-iHt} \tag{16}$$

The fully evolved belief state - a solution of the Schrödinger equation - is obtained by applying the unitary time operator $U(t)$ to the initial state. An inherent feature with Hamiltonian quantum models is the appearance of oscillations of probability over time. In quantum mechanics, finite dimensional and energetically closed systems are always periodical. Therefor the initial belief state will re-occur after the proper time period of the system. In previous work we have argued a third temporal stage, closing the experimental paradigm, should implicitly be supposed in which all driving parameters are set equal to zero [11]. The time of measurement t is taken equal to $\frac{\pi}{2}$ for each stage [14].[3] The final outcome state vector is thus obtained by concatenating both propagators on the initial state, Eq. (5),

$$\Psi_{p|c} = e^{-iH_{p?}\frac{\pi}{2}} e^{-iH_c\frac{\pi}{2}} \Psi_0 \tag{17}$$

The processing of cue c in the first stage occurs till $t = \frac{\pi}{2}$, the processing of probe p? in the second stage takes another time range of $\frac{\pi}{2}$. The time evolution of the acceptance probabilities from initial state processing of cue and processing of probe is shown in Fig. 2.

---

[3] The Hamiltonian parameters are dependent on the choice of measurement time.

Table 1. Observed acceptance ratios and unpacking factors, partitioned by cue type, high frequency & concrete (HFC), high frequency & abstract (HFA), low frequency & concrete (LFC), low frequency & abstract (LFA). N = 70. Data set from Table 2, [8]. Predicted acceptance probabilities and unpacking factors from the Hamiltonian-QEM model, RMSE = 0.054737.

	HFC				HFA				LFC				LFA			
Obs.	L_1	L_2	L_3	L_4	L_1	L_2	L_3	L_4	L_1	L_2	L_3	L_4	L_1	L_2	L_3	L_4
L_1?	0.52	0.31	0.30	0.15	0.52	0.32	0.40	0.25	0.59	0.31	0.34	0.11	0.58	0.44	0.43	0.19
L_2?	0.33	0.35	0.43	0.17	0.36	0.54	0.44	0.24	0.46	0.46	0.35	0.11	0.61	0.53	0.58	0.21
L_3?	0.38	0.35	0.42	0.21	0.37	0.38	0.48	0.24	0.41	0.34	0.49	0.13	0.53	0.38	0.52	0.17
L_{123}?	0.56	0.54	0.60	0.22	0.54	0.64	0.53	0.26	0.64	0.49	0.56	0.13	0.66	0.61	0.59	0.20
UF	2.20	1.89	1.91	2.45	2.32	1.96	2.49	2.83	2.31	2.31	2.10	2.77	2.59	2.19	2.61	2.86
Pred.	L_1	L_2	L_3	L_4	L_1	L_2	L_3	L_4	L_1	L_2	L_3	L_4	L_1	L_2	L_3	L_4
L_1?	0.45	0.36	0.36	0.20	0.49	0.39	0.39	0.19	0.49	0.39	0.39	0.19	0.57	0.47	0.47	0.18
L_2?	0.36	0.45	0.36	0.20	0.39	0.49	0.39	0.19	0.39	0.49	0.39	0.19	0.47	0.57	0.47	0.18
L_3?	0.36	0.36	0.45	0.20	0.39	0.39	0.49	0.19	0.39	0.39	0.49	0.19	0.47	0.47	0.57	0.18
L_{123}?	0.53	0.53	0.53	0.23	0.57	0.57	0.57	0.22	0.57	0.57	0.57	0.22	0.64	0.64	0.64	0.21
UF	2.18	2.18	2.18	2.70	2.24	2.24	2.24	2.65	2.24	2.24	2.24	2.64	2.35	2.35	2.35	2.52

Table 2. Best-fit parameters for Hamiltonian-QEM ($t_1 = t_2 = \pi/2$).

ν	ν'	γ	γ'_{HFC}	γ'_{HFA}	γ'_{LFC}	γ'_{LFA}	κ
-0.6885	0.40345	0.30631	-0.0099825	0.022938	0.027313	0.10107	-0.45978

4   Data and Prediction

We use the experimental 3-list data (N = 70), reported in Brainerd, Wang and Reyna (Table 2, [8]). The CPD model based 'bias-correction' of the acceptance probabilities has been omitted and appear in Table 1. The raw data set was provided by dr. Charles Brainerd. This same data set is used in the CMT model development by Denolf and Lambert-Mogiliansky [15,16], (their Table 1 shows some typos for L_2 cues in the HFA, LFC and LFA set). We optimised the RMSE for 64 data points and Hamiltonian-QEM predictions using 8 parameters $\{\mu, \mu', \gamma, \gamma'_{HFC}, \gamma'_{HFA}, \gamma'_{LFC}, \gamma'_{LFA}, \kappa\}$. Using a 3^8 grid in parameter space the best fit produced RMSE = 0.054737, for the parameter values in Table 2.

5   Discussion

We explored the over distribution predictions of a dynamical extension of Quantum Episodic Memory model by Brainerd, Wang, Reyna and Nakamura [8,9] in the 3-list paradigm. The dynamic processing of the initial belief state in our Hamiltonian-QEM results, over time, in outcome states with adequate values of the acceptance probabilities and the unpacking factor for over distribution of memories. The model further predicts systematic higher acceptance probabilities of targets to their proper source list, $p(L_i?|L_i) > p(L_j?|L_i)$ for $(j \neq i)$, which

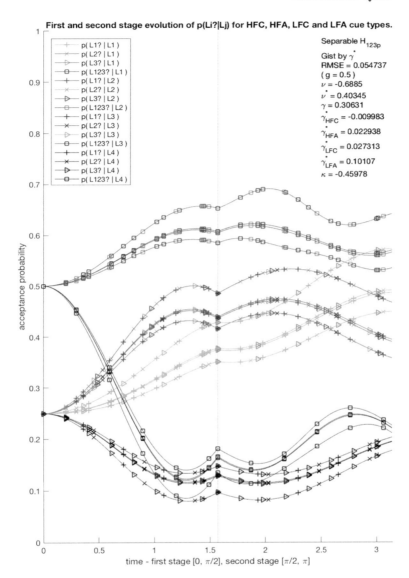

Fig. 2. Optimized two-stage temporal evolution of the acceptance probabilities $p(L_i?|L_j)$ according the Hamiltonian-QEM model with separable H_{123p} for querying the joint-list, and with equal latent gist and verbatim support in the initial state ($g = 0.5$).

is curiously not consistently observed in the experimental data set (N = 70). The model relies on four types of transport embedded in the Hamiltonians; the parameters γ and γ' regulate transport affecting the gist-based component, and the parameters ν and ν' regulate transport affecting the verbatim-based components. The Hamiltonians of the first stage cue recognition phase receive the attenuation parameter κ in the second probe response stage.

The Hamiltonian-QEM model succeeds in a qualitatively good fit with 8 dynamical parameters for the 64 data points of the experimental data set Brainerd et al. [8]. Beyond this exploratory test of the Hamiltonian-QEM for source memory, the model must still be submitted to a comparative statistical test with recent generalisations and modifications of the QEM model, in particular the Generalized-QEM model by Trueblood and Hemmer [25] and Complementary Memory Types model by Denolf and Lambert-Mogiliansky [15, 16, 20].

Acknowledgements. We are grateful to Dr. Charles Brainerd for providing us with the raw data file of the 3-List experiment of [8].

We thank Dr. Peter Kvam for helpful discussions and comments on parts of this manuscript. This research was supported by AFOSR (FA9550-12-1-00397).

References

1. Aerts, D.: Quantum structure in cognition. J. Math. Psychol. **53**, 314–348 (2009). https://doi.org/10.1016/j.jmp.2009.04.005
2. Aerts, D., Broekaert, J., Smets, S.: The Liar-paradox in a quantum mechanical perspective. Found. Sci. **4**(2), 115–132 (1999). https://doi.org/10.1023/A:1009610326206
3. Atmanspacher, H., Filk, T.: A proposed test of temporal nonlocality in bistable perception. J. Math. Psychol. **54**, 314–321 (2010). https://doi.org/10.1016/j.jmp.2009.12.001
4. Brainerd, C.J., Reyna, V.F., Mojardin, A.H.: Conjoint recognition. Psychol. Rev. **106**(1), 160–179 (1999)
5. Brainerd, C.J., Reyna, V.F.: The Science of False Memory. Oxford UP, New York (2005)
6. Brainerd, C.J., Reyna, V.F.: Episodic over-distribution: a signature effect of familiarity without recollection. J. Mem. Lang. **58**, 765–786 (2008)
7. Brainerd, C.J., Reyna, V.F., Holliday, R.E., Nakamura, K.: Overdistribution in source memory. J. Exp. Psychol. Learn. Mem. Cogn. **38**, 413–439 (2012)
8. Brainerd, C.J., Wang, Z., Reyna, V.F.: Superposition of episodic memories: overdistribution and quantum models. Top. Cogn. Sci. **5**, 773–799 (2013)
9. Brainerd, C.J., Wang, Z., Ryena, V.F., Nakamura, K.: Episodic memory does not add up: verbatim-gist superposition predicts violations of the additive law of probability. J. Mem. Lang. **84**, 224–245 (2015)
10. Broekaert, J, Basieva I, Blasiak P, Pothos EM.: Quantum-like dynamics applied to cognition: a consideration of available options. Phil. Trans. R. Soc. A **375** (2017). https://doi.org/10.1098/rsta.2016.0387
11. Broekaert, J., & Busemeyer, J.R.: A Hamiltonian driven quantum like model for overdistribution in episodic memory recollection. Front. Phys. **5** (2017). https://doi.org/10.3389/fphy.2017.00023
12. Bruza, P., Kitto, K., Nelson, D., McEvoy, C.: Is there something quantum-like about the human mental lexicon? J. Math. Psy. **53**(5), 362–377 (2009). https://doi.org/10.1016/j.jmp.2009.04.004

13. Busemeyer, J.R., Pothos, E.M., Franco, R., Trueblood, J.S.: A quantum theoretical explanation for probability judgment errors. Psychol. Rev. **118**(2), 193 (2011). https://doi.org/10.1037/a0022542
14. Busemeyer, J.R., Bruza, P.D.: Quantum Models of Cognition and Decision Making. Cambridge University Press, Cambridge (2012)
15. Denolf, J.: Subadditivity of episodic memory states: a complementarity approach. In: Atmanspacher, H., Bergomi, C., Filk, T., Kitto, K. (eds.) QI 2014. LNCS, vol. 8951, pp. 67–77. Springer, Cham (2015). https://doi.org/10.1007/978-3-319-15931-7_6
16. Denolf, J., Lambert-Mogiliansky, A.: Bohr complementarity in memory retrieval. J. Math. Psychol. **73**, 28–36 (2016)
17. Jacoby, L.L.: A process dissociation framework: separating automatic from intentional uses of memory. J. Mem. Lang. **30**, 513–541 (1991)
18. Kellen, D., Singmann, H., Klauer, K.C.: Modeling source-memory overdistribution. J. Mem. Lang. **76**, 216–236 (2014)
19. Kvam, P.D., Pleskac, T.J., Yu, S., Busemeyer, J.R.: Interference effects of choice on confidence: quantum characteristics of evidence accumulation. PNAS **112**(34), 10645–10650 (2015). https://doi.org/10.1073/pnas.1500688112
20. Lambert-Mogiliansky, A.: Comments on episodic superposition of memory states. Top. Cogn. Sci. **6**, 63–66 (2014). https://doi.org/10.1111/tops.12067
21. Mandler, G.: Recognizing: the judgment of previous occurrence. Psychol. Rev. **87**, 252–271 (1980)
22. Martínez-Martínez, I.: A connection between quantum decision theory and quantum games: the Hamiltonian of strategic interaction. J. Math. Psychol. **58**, 33–44 (2014). https://doi.org/10.1038/srep23812
23. Nelson, D.L., Kitto, K., Galea, D., McEvoy, C.L., Bruza, P.D.: How activation, entanglement, and searching a semantic network contribute to event memory. Mem. Cogn. **41**, 797–819 (2013)
24. Pothos, E.M., Busemeyer, J.R.: Can quantum probability provide a new direction for cognitive modeling? Behav. Brain Sci. **36**, 255–327 (2013). https://doi.org/10.1017/S0140525X12001525
25. Trueblood, J.S., Hemmer, P.: The generalized quantum episodic memory model. Cogn. Sci. **41**, 2089–2125 (2017). https://doi.org/10.1111/cogs.12460
26. Tulving, E.: Precis of elements of episodic memory. Behav. Brain Sci. **7**(2), 223–268 (1984)
27. Wang Z., Busemeyer, J.R.: A quantum question order model supported by empirical tests of an a priori and precise prediction. Top. Cogn. Sci. 1–22 (2013). https://doi.org/10.1111/tops.12040
28. Yearsley, J.M., Pothos, E.M.: Zeno's paradox in decision making. Proc. R. Soc. B **283**, 20160291 (2016). https://doi.org/10.1098/rspb.2016.0291

Decision-Making

Balanced Quantum-Like Model
for Decision Making

Andreas Wichert[1,2(✉)] and Catarina Moreira[1,2]

[1] Department of Computer Science and Engineering,
INESC-ID and Instituto Superior Técnico, Universidade de Lisboa,
Porto Salvo, Portugal
andreas.wichert@tecnico.ulisboa.pt
[2] School of Business, University of Leicester,
University Road, Leicester LE1 7RH, UK
cam74@le.ac.uk

Abstract. Clues from psychology indicate that human cognition is not only based on classical probability theory as explained by Kolmogorov's axioms but additionally on quantum probability. We explore the relation between the law of total probability and its violation resulting in the law of total quantum probability. The violation results from an additional interference that influences the classical probabilities. Outgoing from this exploration we introduce a balanced Bayesian quantum-like model that is based on probability waves. The law of maximum uncertainty indicates how to choose a possible phase value of the wave resulting in a meaningful probability value.

Keywords: Quantum cognition · Law of total probability ·
Probability waves · Decision making

1 Introduction

Clues from psychology indicate that human cognition is not only based on traditional probability theory as explained by Kolmogorov's axioms but additionally on quantum probability [4–8,14]. For example, humans when making decisions violate the law of total probability. The emerging field that studies the corresponding models is called quantum cognition. In this work, we introduce a balanced Bayesian quantum-like model that is based on probability waves. The law of maximum uncertainty indicates how to choose a possible phase value of the wave resulting in a meaningful probability value. We demonstrate the model and the law on several experiments of the literature concerned the prisoner's dilemma game and the two stage gambling game. We compare the results with previous works that deal with predictive quantum-like models for decision making. The results obtained show that the model can make predictions regarding human decision-making with a meaningful interpretation.

© Springer Nature Switzerland AG 2019
B. Coecke and A. Lambert-Mogiliansky (Eds.): QI 2018, LNCS 11690, pp. 79–90, 2019.
https://doi.org/10.1007/978-3-030-35895-2_6

1.1   Prisoner's Dilemma Game and Probability Waves

In the prisoner's dilemma game, there are two prisoners, prisoner x and prisoner y. They have no means of communicating with each other. Each prisoner is offered by the prosecutors a bargain: by testifying against the other one she can betray the other one (Defect). On the other hand, the prisoner can refuse the deal and cooperate with the other one by remaining silent [18].

Several psychological experiments were made assuming that the probability of prisoner x cooperating is $p(x) = 0.5$ and the probability of defecting is $p(\neg x) = 0.5$. The participants of the experiment were asked three different questions.

- What is the probability that the prisoner y defects given x defects, $p(\neg y|\neg x)$.
- What is the probability that the prisoner y defects given x cooperates, $p(\neg y|x)$.
- What is the probability that the prisoner y defects given there is no information present about knowing if prisoner x cooperates or defects. This can be expressed by

$$p(\neg y) = p(\neg y, x) + p(\neg y, \neg x) = p(\neg y|x) \cdot p(x) + p(\neg y|\neg x) \cdot p(\neg x). \quad (1)$$

This relationship can be represented by a graph (see Fig. 1) that indicates the influence between events x and y.

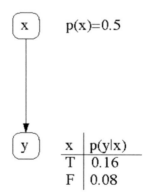

Fig. 1. The causal relation between events x and y represented by a direct graph of two nodes. Note that each node is followed by a conditional probability table that specifies the probability distribution of that node according to its parent node. This direct graph of two nodes representation corresponds to a simple Bayesian network.

In Table 1, we summarise the results of several experiments of the literature concerned with the prisoner's dilemma experiment.

Table 1. Experimental results obtained in four different works of the literature for the prisoner's dilemma game. The column $p(\neg y|\neg x)$ corresponds to the probability of *defecting* given that it is known that the other participant chose to *defect*. The column $p(\neg y|x)$ corresponds to the probability of *defecting* given that it is known that the other participant chose to *cooperate*. Finally, the column $p_{sub}(\neg y)$ corresponds to the subjective probability of the second participant choosing the *defect* action given there is no information present about knowing if prisoner x cooperates or defects. The column $p(\neg y)$ corresponds to the classical probability.

Experiment	$p(\neg y|\neg x)$	$p(\neg y|x)$	$p_{sub}(\neg y)$	$p(\neg y)$
(a) [19]	0.97	0.84	0.63	0.9050
(b) [17]	0.82	0.77	0.72	0.7950
(c) [3]	0.91	0.84	0.66	0.8750
(d) [10]	0.97	0.93	0.88	0.9500
(e) Average	0.92	0.85	0.72	0.8813

1.2   Two Stage Gambling Game

In the two stage gambling game the participants were asked whether they want to play a gamble that has an equal chance of winning $p(x) = 0.5$ or losing $p(\neg x) = 0.5$ [18]. The participants of the experiment were asked three different questions.

- What is the probability that they play the gamble y if they had lost the first gamble x, $p(y|\neg x)$
- What is the probability that they play the gamble y if they had won the first gamble x, $p(y|x)$
- What is the probability that they play the gamble y given there is no information present knowing if they had won the first gamble x. This would by the law of total probability

$$p(y) = p(y|x) \cdot p(x) + p(y|\neg x) \cdot p(\neg x) \tag{2}$$

In Table 2, we summarise the results of several experiments of the literature concerned with the two stage gambling game.

2   Quantum Probabilities and Waves

Beside quantum cognition quantum physics was the only branch in science that evaluated a probability $p(x)$ an state x as the mode-square of a probability amplitude $A(x)$ represented by a complex number

$$p(x) = |A(x)|^2 = \|A(x)\|^2 = A(x)^* \cdot A(x). \tag{3}$$

This is because the product of complex number with its conjugate is always a real number. With

$$A(x) = \alpha + \beta \cdot i \tag{4}$$

Table 2. Experimental results obtained in three different works of the literature indicating the probability of a player choosing to make a second gamble for the two stage gambling game. The column $p(y|\neg x)$ corresponds to the probability when the outcome of the first gamble is known to be lose. The column $p(y|x)$ corresponds to the probability when the outcome of the first gamble is known to be win. Finally, the column $p_{sub}(y)$ corresponds to the subjective probability when the outcome of the first gamble is not known. The column $p(y)$ corresponds to the classical probability.

Experiment	$p(y|\neg x)$	$p(y|x)$	$p_{sub}(y)$	$p(y)$
(i) [20]	0.58	0.69	0.37	0.6350
(ii) [15]	0.47	0.72	0.48	0.5950
(iii) [16]	0.45	0.63	0.41	0.5400
(iv) Average	0.50	0.68	0.42	0.5900

$$A(x)^* \cdot A(x) = (\alpha - \beta \cdot i) \cdot (\alpha + \beta \cdot i) = \alpha^2 + \beta^2 = |A(x)|^2. \tag{5}$$

Quantum physics by itself does not offer any justification or explanation beside the statement that it just works fine [2]. We can map the classical probabilities into amplitudes using the polar coordinate representation

$$a(x, \theta_1) = \sqrt{p(x)} \cdot e^{i \cdot \theta_1} = A(x), \quad a(y, \theta_2) = \sqrt{p(y)} \cdot e^{i \cdot \theta_2} = A(y). \tag{6}$$

The amplitudes represented in polar coordinate form contains a new free parameter θ, which corresponds to the phase of the wave.

2.1 Intensity Waves

The intensity wave is defined as

$$I(y, \theta_1, \theta_2) = |a(y, x, \theta_1) + a(y, \neg x, \theta_2)|^2 \tag{7}$$

$$I(y, \theta_1, \theta_2) = p(y, x) + p(y, \neg x) + 2 \cdot \sqrt{p(y, x) \cdot p(y, \neg x)} \cdot \cos(\theta_1 - \theta_2) \tag{8}$$

Note that for simplification we can replace $\theta_1 - \theta_2$ with θ,

$$\theta = \theta_1 - \theta_2$$

$$I(y, \theta) = p(y) + 2 \cdot \sqrt{p(y, x) \cdot p(y, \neg x)} \cdot \cos(\theta) \tag{9}$$

and

$$I(\neg y, \theta_{\neg 1}, \theta_{\neg 2}) = |a(\neg y, x, \theta_{\neg 1}) + a(\neg y, \neg x, \theta_{\neg 2})|^2 \tag{10}$$

with

$$\theta_\neg = \theta_{\neg 1} - \theta_{\neg 2}$$

$$I(\neg y, \theta_\neg) = p(\neg y) + 2 \cdot \sqrt{p(\neg y, x) \cdot p(\neg y, \neg x)} \cdot \cos(\theta_\neg) \tag{11}$$

and for certain phase values

$$I(y, \theta) + I(\neg y, \theta_\neg) \neq 1. \tag{12}$$

In Fig. 2 (a) we see two intensity waves in relation to the phase with the parametrisation as indicated in corresponding to the values of Fig. 1 corresponding to the Table 1 values in (e).

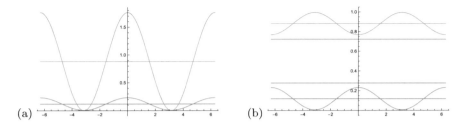

Fig. 2. (a) Two intensity waves $I(y, \theta)$, $I(\neg y, \theta_\neg)$ in relation to the phase $(-2 \cdot \pi, 2 \cdot \pi)$ with the parametrisation as indicated in corresponding to the values of Fig. 1. Note that the two waves oscillate around $p(y) = 0.1950$ and $p(\neg y) = 0.8050$ (the two lines). (b) The resulting probability waves as determined by the law of balance, the bigger wave is replaced by the negative smaller one.

2.2 The Law of Balance and Probability Waves

Intensity waves $I(y, \theta)$ and $I(\neg y, \theta_\neg)$ are probability waves $p(y, \theta)$ and $p(\neg y, \theta_\neg)$ if:

1. they are positive
$$0 \le p(y, \theta), \quad 0 \le p(\neg y, \theta_\neg); \tag{13}$$

2. they sum to one
$$p(y, \theta) + p(\neg y, \theta_\neg) = p(y) + p(\neg y) = 1; \tag{14}$$

3. they are smaller or equal to one
$$p(y, \theta) \le 1, \quad p(\neg y, \theta_\neg) \le 1. \tag{15}$$

Simply speaking, the law states that the bigger wave is replaced by a smaller negative one.

Probability Waves Are Positive. Since the norm is being positive or more precisely non-negative, we can represent a quadratic form by l_2 norm

$$(a(x, \theta_1) + a(y, \theta_2) \cdot a(x, \theta_1) + a(y, \theta_2)) = \|a(x, \theta_1) + a(y, \theta_2)\|^2$$

and it follows
$$0 \le \|a(x, \theta_1) + a(y, \theta_2)\|^2.$$

Probability Waves Sum to One According by the Law of Balance. Instead of simple normalisation of the intensity we propose the law of balance. The interference is balanced, which means that the interference of $p(y, \theta)$ and $p(\neg y, \theta_\neg)$ cancel each out.

$$\sqrt{p(y, x) \cdot p(y, \neg x)} \cdot \cos(\theta) = -\sqrt{p(\neg y, x) \cdot p(\neg y, \neg x)} \cdot \cos(\theta_\neg). \tag{16}$$

We can solve the Equation for the phase $\theta_\neg$ or θ resulting in two possible cases. For $\theta_\neg$ we get

$$\theta_\neg = \cos^{-1}\left(-\sqrt{\frac{p(y|x) \cdot p(y|\neg x)}{p(\neg y|x) \cdot p(\neg y|\neg x)}} \cdot \cos(\theta)\right) \tag{17}$$

Since $\cos^{-1}(x)$ is only defined for $x \in [-1, 1]$ the Eq. 21 is valid for the constraint

$$\frac{p(y|x) \cdot p(y|\neg x)}{p(\neg y|x) \cdot p(\neg y|\neg x)} \leq 1. \tag{18}$$

corresponding to

$$p(y) \leq p(\neg y). \tag{19}$$

Since

$$p(y|x) \cdot p(y|\neg x) \leq p(\neg y|x) \cdot p(\neg y|\neg x)$$
$$p(y|x) \cdot p(y|\neg x) \leq (1 - p(y|x)) \cdot (1 - p(y|\neg x))$$
$$0 \leq 1 - p(y|x) - p(y|\neg x).$$

$p(y) \leq p(\neg y)$ is equivalent for $p(x) = p(\neg x)$ to

$$p(y|x) + p(y|\neg x) \leq (1 - p(y|x)) + (1 - p(y|\neg x))$$
$$0 \leq 1 - p(y|x) - p(y|\neg x).$$

The smaller probability wave determines the other probability wave as

$$p(\neg y, \theta_\neg) = 1 - p(y, \theta). \tag{20}$$

If the preceding constraint is not valid, then we solve the equation for the phase θ. It has to be valid with

$$\theta = \cos^{-1}\left(-\sqrt{\frac{p(\neg y|x) \cdot p(\neg y|\neg x)}{p(y|x) \cdot p(y|\neg x)}} \cdot \cos(\theta_\neg)\right) \tag{21}$$

and the constraint

$$p(\neg y) \leq p(y) \tag{22}$$

simplified as

$$p(y, \theta) = 1 - p(\neg y, \theta_\neg). \tag{23}$$

For equality in the constraint $p(y) = p(\neg y)$ both cases become equal.

Probability Waves Are Smaller Equal One. We assume without loss of generality

$$p(y) \leq p(\neg y) \tag{24}$$

it follows that

$$p(y) \leq 0.5. \tag{25}$$

By the inequality of arithmetic and geometric means

$$\sqrt{p(y,x) \cdot p(y,\neg x)} \leq \frac{p(y,x) + p(y,\neg x)}{2} \tag{26}$$

$$2 \cdot \sqrt{p(y,x) \cdot p(y,\neg x)} \leq p(y) = p(y,x) + p(y,\neg x) \tag{27}$$

$$p(y,\theta) = p(y) + 2 \cdot \sqrt{p(y|x) \cdot p(x) \cdot p(y|\neg x) \cdot p(\neg x)} \cdot \cos(\theta) \leq 2 \cdot p(y) \leq 1. \tag{28}$$

2.3   Probability Waves

Using the values of Tables 1 and 2, we can determine the probability waves

$$p(\neg y, \theta_\neg) = p(\neg y) + 2 \cdot \sqrt{p(\neg y|x) \cdot p(x) \cdot p(\neg y|\neg x) \cdot p(\neg x)} \cdot \cos(\theta_\neg). \tag{29}$$

and

$$p(y,\theta) = p(y) + 2 \cdot \sqrt{p(y|x) \cdot p(x) \cdot p(y|\neg x) \cdot p(\neg x)} \cdot \cos(\theta). \tag{30}$$

as indicated in Fig. 3. For

$$p(y) \leq p(\neg y) \tag{31}$$

the maximal interference is

$$\pm Inter_{max} = \pm\sqrt{p(y,x) \cdot p(y,\neg x)} \tag{32}$$

and for

$$p(\neg y) \leq p(y) \tag{33}$$

the maximal interference is

$$\pm Inter_{max} = \pm\sqrt{p(\neg y,x) \cdot p(\neg y,\neg x)}. \tag{34}$$

We can define the intervals that describe the probability waves as

$$I_y = [p(y) - Inter_{max}, p(y) + Inter_{max}] \tag{35}$$

and

$$I_{\neg y} = [p(\neg y) - Inter_{max}, p(\neg y) + Inter_{max}] \tag{36}$$

with

$$p(\neg y, \theta_\neg) \in I_{\neg y}, \quad p(y,\theta) \in I_y \tag{37}$$

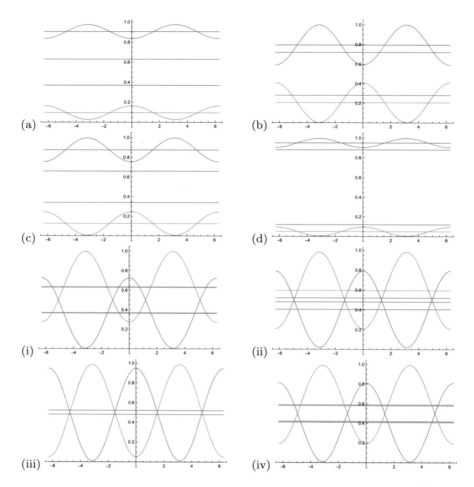

Fig. 3. Probability waves for the experiments described in Tables 1 and 2. In plots (a) - (d) the waves $p(\neg y, \theta_\neg)$ are around $p(\neg y)$, (for (e) see Fig. 2). In the plots (i) - (iv) the waves $p(y, \theta)$ are around $p(y)$. Additionally the values $p_{sub}(\neg y)$ and $p_{sub}(y)$ are indicated by a line. Note that the curves in the plots (i) – (iv) overlap.

3   Law of Maximal Uncertainty

How can we choose the phase θ for a probability value that reflects the case of not knowing what the correct value is and without losing information about the probability waves? We answer this question by proposing the law of maximal uncertainty is based on two principles, the principle of entropy and the mirror principle.

3.1 Principle of Entropy

For the case in which both interval do not overlap

$$I_y \cap I_{\neg y} = \emptyset \tag{38}$$

the values of the waves that are closest to the equal distribution are chosen. By doing so the uncertainty is maximised and the information about the probability wave is not lost. The principle of maximum entropy states that the probability distribution which best represents the current state of knowledge is the one with largest entropy [11–13]. In the case of a binary event the highest entropy corresponds to an equal distribution

$$H = -p(y) \cdot \log_2 p(y) - p(\neg y) \cdot \log_2 p(\neg y) = -\log_2 \cdot 0.5 = 1 \; bit. \tag{39}$$

For $p(y) \leq p(\neg y)$ the closest values to the equal distribution are for $\theta = 0$, the ends of the interval for

$$p_q(y) = p(y) + 2 \cdot \sqrt{p(y, x) \cdot p(y, \neg x)} \approx 2 \cdot p(y) \tag{40}$$

and

$$p_q(\neg y) = 1 - p_q(y). \tag{41}$$

For $p(\neg y) \leq p(y)$ the closest values to the equal distribution are

$$p(\neg y)_q = p(\neg y) + 2 \cdot \sqrt{p(\neg y, x) \cdot p(\neg y, \neg x)} \approx 2 \cdot p(\neg y) \tag{42}$$

and

$$p_q(y) = 1 - p_q(\neg y). \tag{43}$$

3.2 Mirror Principle

For the case where the intervals overlap

$$I_y \cap I_{\neg y} \neq \emptyset \tag{44}$$

an equal distribution maximises the uncertainty but loses the information about the probability wave. To avoid the loss we do not change the entropy of the system we use the positive interference as defined by the law of balance. When the intervals overlap the positive interference is approximately of the size of smaller probability value since the arithmetic and geometric means approach each other, see Eq. 26. We increase uncertainty by mirror the "probability values". For the case $p(y) \leq p(\neg y)$ we assume

$$p_q(\neg y) = 2 \cdot \sqrt{p(y, x) \cdot p(y, \neg x)} \approx p(y) \tag{45}$$

and

$$p_q(y) = 1 - p_q(\neg y). \tag{46}$$

For $p(\neg y) \le p(\neg y)$

$$p_q(y) = 2 \cdot \sqrt{p(\neg y, x) \cdot p(\neg y, \neg x)} \approx p(\neg y) \qquad (47)$$

and

$$p_q(\neg y) = 1 - p_q(y). \qquad (48)$$

Table 3 summarises the intervals that describe the probability waves, the resulting probabilities p_q that are based on the law of maximal uncertainty, the subjective probability and the classical probability values. We compare the results that

Table 3. Probability waves, the resulting probabilities p_q that are based on the law of maximal uncertainty, the subjective probability and the classical probability values. Entries (a)–(e) are based on the principle of entropy and entries (i)–(iv) are based mirror principle.

Experiment	$I_{\neg y}$	$p_{sub}(\neg y)$	$p_q(\neg y)$	$p(\neg y)$
(a)	[0.84, 0.97]	0.63	0.84	0.91
(b)	[0.59, 1]	0.72	0.59	0.79
(c)	[0.76, 1]	0.66	076	0.88
(d)	[0.90, 1]	0.88	0.90	0.95
(e) Average	[0.77, 0.99]	0.72	0.77	0.88
Experiment	I_y	$p_{sub}(y)$	$p_q(y)$	$p(y)$
(i)	[0.27, 0.98]	0.37	0.36	0.64
(ii)	[0.20, 0.98]	0.48	0.39	0.59
(iii)	[0.09, 0.99]	0.41	0.45	0.54
(vi) Average	[0.19, 0.99]	0.42	0.40	0.59

are based on probability waves and the law of maximal uncertainty with previous works that deal with predictive quantum-like models for decision making, see Table 4. The dynamic heuristic [18] is used quantum-like bayesian networks. Its parameters are determined by examples from a domain. On the other hand In the Quantum Prospect Decision Theory the values need not to be adapted to a domain. The quantum interference term is determined by the Interference Quarter Law. The quantum interference term of total probability is simply fixed to a value equal to 0.25 [22].

Table 4. Comparison between the Quantum Prospect Decision Theory (DT) [22], the dynamic heuristic (DH) [18] and the law of maximal uncertainty (MU) of the balanced quantum-like model. The results of the dynamic heuristic (DH) and the law of maximal uncertainty (MU) are similar, however the law of maximal uncertainty (MU) was not adapted to a domain.

Experiment	*observed*	*PDT*	*DH*	*MU*
(a)	0.63	0.65	0.64	0.84
(b)	0.72	0.54	0.71	0.59
(c)	0.66	0.63	0.80	0.76
(d)	0.88	0.70	0.90	0.90
(e) Average	0.72	0.63	0.76	0.77
(i)	0.37	0.39	0.36	0.36
(ii)	0.48	0.35	0.40	0.39
(iii)	0.41	0.29	0.41	0.45
(iv) Average	0.42	0.34	0.39	0.40

4   Conclusion

Physical experiments indicate that wave functions are present in the world [21]. They state that the size does not matter and that a very large number of atoms can be entangled [1,9]. Clues from psychology also indicate that human cognition is based on quantum probability rather than the traditional probability theory as explained by Kolmogorov's axioms [5–8]. This approach could lead to the conclusion that a wave function can be present at the macro scale of our daily life.

We introduce a balanced Bayesian quantum-like model that is based on probability waves. The law of maximum uncertainty indicates how to choose a possible phase value of the wave resulting in a meaningful probability value. The law of maximal uncertainty of the balanced quantum-like model is not static, meaningful and does not need to be adapted to a specific domain. The results obtained show that the model can make predictions regarding human decision-making with a meaningful interpretation.

Acknowledgment. This work was supported by national funds through Fundação para a Ciência e a Tecnologia (FCT) with reference UID/CEC/50021/2013. The funders had no role in study design, data collection and analysis, decision to publish, or preparation of the manuscript.

References

1. Amico, L., Fazio, R., Osterloh, A., Vedral, V.: Entanglement in many-body systems. Rev. Mod. Phys. **80**(2), 517–576 (2008)
2. Binney, J., Skinner, D.: The Physics of Quantum Mechanics. Oxford University Press, Oxford (2014)

3. Busemeyer, J., Matthew, M., Wang, Z.: A quantum information processing explanation of disjunction effects. In: Proceedings of the 28th Annual Conference of the Cognitive Science Society, pp. 131–135 (2006)
4. Busemeyer, J., Wang, Z.: Quantum cognition: key issues and discussion. Top. Cogn. Sci. **6**, 43–46 (2014)
5. Busemeyer, J.R., Bruza, P.D.: Quantum Models of Cognition and Decision. Cambridge University Press, Cambridge (2012)
6. Busemeyer, J.R., Trueblood, J.: Comparison of quantum and Bayesian inference models. In: Bruza, P., Sofge, D., Lawless, W., van Rijsbergen, K., Klusch, M. (eds.) QI 2009. LNCS (LNAI), vol. 5494, pp. 29–43. Springer, Heidelberg (2009). https://doi.org/10.1007/978-3-642-00834-4_5
7. Busemeyer, J.R., Wang, Z., Lambert-Mogiliansky, A.: Empirical comparison of Markov and quantum models of decision making. J. Math. Psychol. **53**(5), 423–433 (2009). https://doi.org/10.1016/j.jmp.2009.03.002
8. Busemeyer, J.R., Wang, Z., Townsend, J.T.: Quantum dynamics of human decision-making. J. Math. Psychol. **50**(3), 220–241 (2006). https://doi.org/10.1016/j.jmp.2006.01.003
9. Ghosh, S., Rosenbaum, T.F., Aeppli, G., Coppersmith, S.N.: Entangled quantum state of magnetic dipoles. Nature **425**, 48–51 (2003)
10. Hristova, E., Grinberg, M.: Disjunction effect in prisoner's dilemma: evidences from an eye-tracking study. In: Proceedings of the 30th Annual Conference of the Cognitive Science Society, pp. 1225–1230 (2008)
11. Jaynes, E.T.: Information theory and statistical mechanics. Phys. Rev. Ser. II **106**(4), 620–630 (1957)
12. Jaynes, E.T.: Information theory and statistical mechanics II. Phys. Rev. Ser. II **108**(2), 171–190 (1957)
13. Jaynes, E.T.: Prior probabilities. IEEE Trans. Syst. Sci. Cybern. **4**(3), 227–241 (1968)
14. Khrennikov, A.: Quantum-like model of cognitive decision making and information processing. J. BioSyst. **95**, 179–187 (2009)
15. Kuhberger, A., Komunska, D., Josef, P.: The disjunction effect: does it exist for two-step gambles? Organ. Behav. Hum. Decis. Process. **85**, 250–264 (2001)
16. Lambdin, C., Burdsal, C.: The disjunction effect reexamined: relevant methodological issues and the fallacy of unspecified percentage comparisons. Organ. Behav. Hum. Decis. Process. **103**, 268–276 (2007)
17. Li, S., Taplin, J.: Examining whether there is a disjunction effect in prisoner's dilemma game. Chin. J. Psychol. **44**, 25–46 (2002)
18. Moreira, C., Wichert, A.: Quantum-like Bayesian networks for modeling decision making. Front. Psychol. **7**, 11 (2016)
19. Shafir, E., Tversky, A.: Thinking through uncertainty: nonconsequential reasoning and choice. Cogn. Psychol. **24**, 449–474 (1992)
20. Tversky, A., Shafir, E.: The disjunction effect in choice under uncertainty. J. Psychol. Sci. **3**, 305–309 (1992)
21. Vedral, V.: Living in a quantum world. Sci. Am. **304**(6), 38–43 (2011)
22. Yukalov, V., Sornette, D.: Decision theory with prospect interference and entanglement. Theor. Decis. **70**, 283–328 (2011)

Introducing Quantum-Like Influence Diagrams for Violations of the Sure Thing Principle

Catarina Moreira[1(✉)] and Andreas Wichert[2]

[1] School of Business, University of Leicester,
University Road, Leicester LE1 7RH, UK
cam74@le.ac.uk
[2] Instituto Superior Técnico, INESC-ID,
Av. Professor Cavaco Silva, 2744-016 Porto Salvo, Portugal
andreas.wichert@tecnico.ulisboa.pt

Abstract. It is the focus of this work to extend and study the previously proposed quantum-like Bayesian networks (Moreira and Wichert, 2014, 2016) to deal with decision-making scenarios by incorporating the notion of maximum expected utility in influence diagrams. The general idea is to take advantage of the quantum interference terms produced in the quantum-like Bayesian Network to influence the probabilities used to compute the expected utility of some action. This way, we are not proposing a new type of expected utility hypothesis. On the contrary, we are keeping it under its classical definition. We are only incorporating it as an extension of a probabilistic graphical model in a compact graphical representation called an influence diagram in which the utility function depends on the probabilistic influences of the quantum-like Bayesian network.

Our findings suggest that the proposed quantum-like influence diagram can indeed take advantage of the quantum interference effects of quantum-like Bayesian Networks to maximise the utility of a cooperative behaviour in detriment of a fully rational defect behaviour under the prisoner's dilemma game.

Keywords: Quantum cognition · Quantum-like influence diagrams · Quantum-Like Bayesian Networks

1 Introduction

In this work, we extend the Quantum-Like Bayesian Network previously proposed by Moreira and Wichert (2014, 2016) by incorporating the framework of expected utility. This extension is motivated by the fact that quantum-like models tend to explain the probability distributions in several decision scenarios where the agent (or the decision-maker) tends to act irrationally (Busemeyer and Bruza 2012; Bruza et al. 2015). By irrational, we mean that an individual

© Springer Nature Switzerland AG 2019
B. Coecke and A. Lambert-Mogiliansky (Eds.): QI 2018, LNCS 11690, pp. 91–108, 2019.
https://doi.org/10.1007/978-3-030-35895-2_7

chooses strategies that do not maximise or violate the axioms of expected utility. It is not enough to know these probability distributions. On the contrary, it would be desirable to use this probabilistic information to help us act upon a real world decision scenario. For instance, if a patient has cancer, it is not enough for a doctor to know the probability distribution of success of different treatments. The doctor needs to act and choose a treatment based on specific information about the patient and how this treatment will affect him/her. Decision-making models such as the expected utility hypothesis are used to decide how to act in the world. The main problem with such decision-making models is that it is very challenging to determine the right action in a decision task where the outcomes of the actions are not fully determined (Koller and Friedman 2009). For this reason, we suggest to extend the previously proposed Quantum-Like Bayesian Network to a Quantum-Like Influence diagram where we take into account both the quantum-like probabilities (incorporating quantum interference effects) of the various outcomes and the preferences of an individual between these outcomes.

Generally speaking, an Influence diagram is a compact directed acyclical graphical representation of a decision scenario originally proposed by Howard and Matheson (1984) which consists in three types of nodes: random variables (nodes) of a Bayesian Network, action nodes representing a decision that we need to make, and an utility function. The goal is to make a decision, which maximises the expected utility function by taking into account probabilistic inferences performed on the Bayesian Network. However, since influence diagrams are based on classical Bayesian Networks, then they cannot cope with the paradoxical findings reported over the literature.

It is the focus of this work to study the implications of incorporating Quantum-Like Bayesian Networks in the context of influence graphs. By doing so, we are introducing quantum interference effects that can disturb the final probability outcomes of a set of actions and affect the final expected utility. We will study how one can use influence diagrams to explain the paradoxical findings of the prisoner's dilemma game based on expected utilities.

2 Revisiting the Prisoner's Dilemma and the Expected Utility Hypothesis

The Prisoner's Dilemma game consists in two players who are in two separate confinements with no means of communicating with each other. They were offered a deal: if one defects against the other, he is set free while the other gets a heavy charge. If they both defect, they get both a big charge and if they both cooperate by remaining silent, they get a small charge. Figure 1 shows an example of a payoff matrix for the Prisoner's Dilemma used in the experiments of Shafir and Tversky (1992) where the goal is to score the maximum number of points.

Looking at the payoff matrix, one can see that the best action for *both* players is to *cooperate*, however experimental findings show that the majority of the

| | Player 2 | |
	Defect	Cooperate
Defect	P1: 30 P2: 30	P1: 85 P2: 25
Cooperate	P1: 25 P2: 85	P1: 75 P2: 75

(Player 1 labels the two data rows)

Fig. 1. Example of a payoff matrix used in the Shafir and Tversky (1992) Prisoner's Dilemma experiment

players choose to *defect* even when it is known that the other player chose to *cooperate*. The Prisoner's Dilemma is a clear example of how two perfectly rational individuals choose to defect (they prefer an individual reward), rather than choosing the option which is best for both (to cooperate). The expected utility hypothesis is a framework that enables us to explain why this happens.

The expected utility hypothesis corresponds to a function designed to take into account decisions under risk. It consists of a choice of a possible set of actions represented by a probability distribution over a set of possible payoffs (von Neumann and Morgenstern 1953). It is given by Eq. 1,

$$EU = \sum_i Pr(x_i) \cdot U(x_i), \tag{1}$$

where $U(x_i)$ is an utility function associated to event x_i.

In the experiment of Shafir and Tversky (1992), the participant needed to choose between de actions *defect* or *cooperate*. We will address to this participant as player 2, $P2$, and his opponent, to player 1, $P1$. According to the expected utility hypothesis, $P2$ would have to choose the action that would grant him the highest expected utility. Assuming that we do not know what $P1$ chose (so we model this with a neutral prior of 0.5), we can compute the expected utility of Player 2 as

$$EU[Defect] = 0.5 \times U(P1 = D, P2 = D) + 0.5 \times U(P1 = C, P2 = D) = 57.5,$$

$$EU[Cooperate] = 0.5 \times U(P1 = D, P2 = C) + 0.5 \times U(P1 = C, P2 = C) = 50.$$

Note that $U(P1 = x, P2 = y)$ corresponds to the utility of player 1 choosing action x and player 2 choosing action y. The calculations show that the action that maximises the player's expected utility is *Defect*. This is what it is known as the *Maximum Expected Utility hypothesis* (MEU).

In the end of the 70's, Daniel Kahneman and Amos Tversky showed in a set of experiments that in many real life situations, the predictions of the expected utility were completely inaccurate (Tversky and Kahneman 1974; Kahneman et al. 1982; Kahneman and Tversky 1979). This means that a decision theory should be predictive in the sense that it should say what people actually *do choose*, instead of what they *must choose*. The Prisoner's Dilemma game is one of the experiments that show the inaccuracy of the expected utility hypothesis by showing violations to the laws of classical probability and to the Sure Thing Principle. Table 1 summarises the results of several works of the literature reporting violations to the Sure Thing Principle. All of these works tested three conditions in the Prisoners Dilemma Game: (1) the player knows the other defected (*Known to Defect*), (2) the player knows the other cooperated (*Known to Collaborate*), (3) the player does not know the other player's action (*Unknown*). This last condition shows a deviation from the classical probability theory, suggesting that there is a significant percentage of players who are not acting according to the maximum expected utility hypothesis. The Sure Thing Principle (Savage 1954) principle is fundamental in the Bayesian probability theory and states that if one prefers action A over B under state of the world X, and if one also prefers A over B under the complementary state of the world X, then one should always prefer action A over B even when the state of the world is unspecified. Violations of the Sure Thing Principle imply violations of the classical law of total probability.

Table 1. Works of the literature reporting the probability of a player choosing to defect under several conditions. The entries of the table that are highlighted correspond to experiments where the violations of the sure thing principle were not found.

Literature	Known to defect	Known to collaborate	Unknown	Classical probability
Shafir and Tversky (1992)	0.9700	0.8400	0.6300	0.9050
Li and Taplin (2002) (Average)	0.8200	0.7700	0.7200	0.7950
Li and Taplin (2002) **Game 1**	0.7333	0.6670	0.6000	0.7000
Li and Taplin (2002) **Game 2**	0.8000	0.7667	0.6300	0.7833
Li and Taplin (2002) **Game 3**	**0.9000**	**0.8667**	**0.8667**	**0.8834**
Li and Taplin (2002) **Game 4**	0.8333	0.8000	0.7000	0.8167
Li and Taplin (2002) **Game 5**	0.8333	0.7333	0.7000	0.7833
Li and Taplin (2002) **Game 6**	**0.7667**	**0.8333**	**0.8000**	**0.8000**
Li and Taplin (2002) **Game 7**	**0.8667**	**0.7333**	**0.7667**	**0.8000**

Table 1 presents several examples where the principle of maximum expected utility is not, in general, an adequate descriptive model of human behaviour. In fact, people are often irrational, in the sense that their choices do not satisfy the principe of maximum expected utility relative to any utility function (Koller and Friedman 2009).

Previous works in the literature have proposed quantum-like probabilistic models that try to accommodate these paradoxical scenarios and violations to the Sure Thing Principle (Busemeyer et al. 2006b, 2009; Pothos and Busemeyer 2009; Busemeyer and Bruza 2012). There is also a vast amount of work in trying to extend the expected utility hypothesis to a quantum-like versions Mura (2009); Yukalov and Sornette (2015). However, the expected utility framework alone poses some difficulties, since it is very challenging the task of decision-making in situations where the outcomes of an action are not fully determined (Koller and Friedman 2009).

In this paper, we try to fill this gap by taking into account the quantum-like probability inferences produced by a quantum-like Bayesian network to various outcomes and extend these probabilities to influence the preferences of an individual between these outcomes. Note that the probabilistic inferences produced by the quantum-like Bayesian network will suffer quantum interference effects in decision scenarios under uncertainty. The general idea is to use these quantum interference effects to influence the expected utility framework in order to favour other actions than what would be predicted from the classical theory alone. We will combine this structure in a directed and acyclic compact probabilistic graphical model for decision-making, which we will define as the quantum-like influence diagram.

3   A Quantum-Like Influence Diagram for Decision-Making

A Quantum-Like Influence Diagram is a compact directed acyclical graphical representation of a decision scenario, which was originally proposed by Howard and Matheson (1984). It consists on a set of random variables $X_1, \ldots, X_N$ belonging to a quantum-like Bayesian network. Each random variable X_i is associated with a conditional probability distribution (CPD) table, which describes the distribution of quantum probability amplitudes of the random variable X_i with respect to its parent nodes, $\psi(X_i|Pa_{X_i})$. Note that the difference between a quantum-like Bayesian network and a classical network is simply the usage of complex numbers instead of classical real numbers. The usage of complex numbers will enable the emergence of quantum interference effects. The influence diagram also consists in an utility node defined variable U, which is associated with a deterministic function $U(Pa_U)$. The goal is to make a decision, which maximises the expected utility function by taking into account probabilistic inferences performed on the quantum-like Bayesian network.

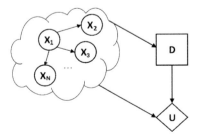

Fig. 2. General example of a Quantum-Like Influence Diagram comprised of a Quantum-Like Bayesian Network, $X_1, ..., X_N$, a Decision Node, D, and an Utility node with no children, U.

An example of a quantum-like influence diagram is presented in Fig. 2. One can notice the three different types of nodes: (1) random variable nodes (circle-shape), denoted by $X_1, \cdots, X_N$, of some Quantum-Like Bayesian Network, (2) a decision node (rectangle-shape), denoted by D, which corresponds to the decision that we want to make, and (3) an Utility node (diamond-shape), denoted by U, which in the scope of this paper, will represent the payoffs in the Prisoner's Dilemma Game.

The goal is to maximise the expected utility by taking into consideration the probabilistic inferences of the quantum-like Bayesian Network, which makes use of the quantum interference effects to accommodate and predict violations to the Sure Thing Principle.

In the next sections, we will address each of these three components separately.

4   Quantum-Like Bayesian Networks

Quantum-like Bayesian Network have been initially proposed by Moreira and Wichert (2014, 2016) and they can be defined by a directed acyclic graph structure in which each node represents a different quantum random variable and each edge represents a direct influence from the source node to the target node. The graph can represent independence relationships between variables, and each node is associated with a conditional probability table that specifies a distribution of quantum complex probability amplitudes over the values of a node given each possible joint assignment of values of its parents. In other words, a quantum-like Bayesian Network is defined in the same way as classical network with the difference that real probability values are replaced by complex probability amplitudes.

In order to perform exact inferences in a quantum-like Bayesian network, one needs to compute the:

- **Quantum-Like full join probability distribution.** The quantum-like full joint complex probability amplitude distribution over a set of N random variables $\psi(X_1, X_2, ..., X_N)$ corresponds to the probability distribution assigned to all of these random variables occurring together in a Hilbert space. Then, the full joint complex probability amplitude distribution of a quantum-like Bayesian Network is given by:

$$\psi(X_1, \ldots, X_N) = \prod_{j=1}^{N} \psi(X_j | Parents(X_j)) \qquad (2)$$

Note that, in Eq. 2, X_i is the list of random variables (or nodes of the network), $Parents(X_i)$ corresponds to all parent nodes of X_i and $\psi(X_i)$ is the complex probability amplitude associated with the random variable X_i. The probability value is extracted by applying Born's rule, that is, by making the squared magnitude of the joint probability amplitude, $\psi(X_1, \ldots, X_N)$:

$$Pr(X_1, \ldots, X_N) = |\psi(X_1, \ldots, X_N)|^2 \qquad (3)$$

– **Quantum-Like Marginalization.** Given a query random variable X and let Y be the unobserved variables in the network, the marginal distribution of X is simply the amplitude probability distribution of X averaging over the information about Y. The quantum-like marginal probability can be defined by Eq. 4. The summation is over all possible y, i.e., all possible combinations of values of the unobserved values y of variable Y. The term γ corresponds to a normalisation factor. Since the conditional probability tables used in Bayesian Networks are not unitary operators with the constraint of double stochasticity (like it is required in other works of the literature (Busemeyer et al. 2006b; Pothos and Busemeyer 2009)), we need to normalise the final scores. This normalisation is consistent with the notion of normalisation of wave functions used in Feynman's Path Diagrams.

In classical Bayesian inference, on the other hand, normalisation is performed due to the independence assumptions made in Bayes rule.

$$Pr(X|e) = \gamma \left| \sum_{y} \prod_{k=1}^{N} \psi(X_k|Parents(X_k), e, y) \right|^2 \tag{4}$$

Expanding Eq. 4, it will lead to the quantum marginalisation formula (Moreira and Wichert 2014), which is composed by two parts: one representing the classical probability and the other representing the quantum interference term (which corresponds to the emergence of destructive / constructive interference effects):

$$Pr(X|e) = \gamma \sum_{i=1}^{|Y|} \left| \prod_{k}^{N} \psi(X_k|Parents(X_k), e, y=i) \right|^2 + 2 \cdot Interference \tag{5}$$

$$Interference =$$
$$\sum_{i=1}^{|Y|-1} \sum_{j=i+1}^{|Y|} \left| \prod_{k}^{N} \psi(X_k|Parents(X_k), e, y=i) \right| \cdot \left| \prod_{k}^{N} \psi(X_k|Parents(X_k), e, y=j) \right| \cdot \cos(\theta_i - \theta_j)$$

Note that, in Eq. 5, if one sets $(\theta_i - \theta_j)$ to $\pi/2$, then $\cos(\theta_i - \theta_j) = 0$. This means that the quantum interference term is canceled and the quantum-like Bayesian Network collapses to its classical counterpart.

Formal methods to assign values to quantum interference terms are still an open research question, however some work has already been done towards that direction (Yukalov and Sornette 2011; Moreira and Wichert 2016, 2017).

5   Maximum Expected Utility in Classical Influence Diagrams

In a decision scenario, D, given a set of possible decision rules, δ_A, the goal of Influence Diagrams is to compute the decision rule that leads to the Maximum Expected Utility. Additionally, $Pr_{\delta_A}(x|a)$ corresponds to a full joint probability distribution of all possible outcomes, x, given different actions a belonging to the decision rules δ_A.

$$EU\left[\mathcal{D}\left[\delta_A\right]\right] = \sum_x Pr_{\delta_A}\left(x|a\right)U\left(x,a\right) \tag{6}$$

The goal is to choose some action a that maximises the expected utility with respect to some decision rule, δ_A:

$$a^* = argmax_{\delta_A} EU\left[\mathcal{D}\left[\delta_A\right]\right]$$

One can map the expected utility formalism to the scope of Bayesian networks in the following way. Knowing that $Pr_{\delta_A}(x|a)$ corresponds to a full joint probability distribution of all possible outcomes, x, given different actions a belonging to the decision rules δ_A, this means that we can decompose the full joint probability distribution to the chain rule of probability theory as the product of each node with its parent nodes.

$$EU\left[\mathcal{D}\left[\delta_A\right]\right] = \sum_x Pr_{\delta_A}\left(x|a\right)U\left(x,a\right) \tag{7}$$

$$EU\left[\mathcal{D}\left[\delta_A\right]\right] = \sum_{X_1,\ldots,X_n,A}\left(\left(\prod_i Pr\left(X_i|Pa_{X_i}\right)\right)U\left(Pa_U\right)\delta_A\left(A|Z\right)\right) \tag{8}$$

In Eq. 8, $Z = Pa_A$ represents the parent nodes of action A. We can factorise Eq. 8 in terms of the decision rule, δ_A, obtaining

$$EU\left[\mathcal{D}\left[\delta_A\right]\right] = \sum_{Z,A}\delta_A\left(A|Z\right)\left(\sum_W\left(\prod_i Pr\left(X_i|Pa_{X_i}\right)\right)U\left(Pa_U\right)\right), \tag{9}$$

where $W = \{X_1,\ldots,X_N\} - Z$ corresponds to all nodes of the Bayesian Network that are not contained in the set of nodes in Z.

By marginalising the summation over W, we obtain an expected utility formula that is written only in terms of the factor $\mu(A, Z)$. Note that this factor corresponds to a conditional distribution table of random variable Z (the outcomes of some action a) and action a.

$$EU\left[\mathcal{D}\left[\delta_A\right]\right] = \sum_{Z,A}\delta_A\left(A|Z\right)\mu\left(A,Z\right) \tag{10}$$

The Maximum Expected Utility for a classical Influence Diagrams is given by (Koller and Friedman 2009):

$$\delta_A^*\left(a,Z\right) = \alpha(x) = \begin{cases} 1 & a = argmax\left(A,Z\right) \\ 0 & otherwise \end{cases} \tag{11}$$

6 Maximum Expected Utility in Quantum-Like Influence Diagrams

The proposed quantum-like influence diagram is built upon the formalisms of quantum-like Bayesian networks. This means that real classical probabilities need to be replaced by complex quantum amplitudes.

We start the derivation with the initial notion of expected utility already presented in the previous section. In a decision scenario, D, given a set of possible decision rules, δ_A, the goal of the Quantum-Like Influence Diagrams is to compute the decision rule that leads to the Maximum Expected Utility. $Pr_{\delta_A}(x|a)$ corresponds to a full joint probability distribution of all possible outcomes, x, given different actions a belonging to the decision rules δ_A.

$$EU\left[\mathcal{D}\left[\delta_A\right]\right] = \sum_x Pr_{\delta_A}\left(x|a\right)U\left(x,a\right) \tag{12}$$

For simplicity, let's consider the decision scenario where we have two binary events X_1 and X_2. Then, we can decompose the classical expected utility equation as

$$EU\left[\mathcal{D}\left[\delta_A\right]\right] = \sum_{X_1,X_2,A} \delta_A(A|X_1)Pr\left(X_1\right)Pr\left(X_2|X_1\right)U\left(X_1,A\right) \tag{13}$$

Like before, we can factorise this formula in terms of the decision rule δ_A, obtaining

$$EU\left[\mathcal{D}\left[\delta_A\right]\right] = \sum_{A,X_2} \delta_A(A|X_2)\sum_{X_1} Pr\left(X_1\right)Pr\left(X_2|X_1\right)U\left(X_1,A\right) \tag{14}$$

For binary events, we obtain the marginalisation of X_1 over both X_2 and A

$$EU\left[\mathcal{D}\left[\delta_A\right]\right] = \sum_{A,X_2} \delta_A(A|X_2)\cdot\mu\left(X_2,A\right) \tag{15}$$

where $\mu\left(X_2,A\right)$ is a factor with the utility function expressed in terms of the distribution of X_2. More specifically, it is given by

$$\begin{aligned} \mu\left(X_2,A\right) = Pr\left(X_1=t\right)Pr\left(X_2|X_1=t\right)U\left(X_1=t,A\right) \\ +Pr\left(X_1=f\right)Pr\left(X_2|X_1=f\right)U\left(X_1=f,A\right) \end{aligned} \tag{16}$$

Since the proposed quantum-like influence diagram makes use of a quantum-like Bayesian network, this means that we need to convert the classical real probabilities into complex quantum amplitudes. This is performed by applying Born's rule: for some classical probability A, the corresponding quantum amplitude is simply its squared magnitude, $Pr(A) = |\psi_A|^2$ (Deutsch 1988; Zurek 2011). Since in Eq. 16 we have an utility factor expressed in terms of the probability distribution of X_1, we cannot apply Born's rule directly, since we would not be satisfying its definition. For this reason, it is not possible to make a direct mapping between this joint probability distribution over an utility function into a quantum-like scenario.

We propose to split Eq. 16 into two factors: one containing a classical probability and another containing the utility function. This procedure is inspired the Quantum Decision Theory model of Yukalov and Sornette (2015). In their model, a prospect π_a is a composite event represented in the Hilbert space by an eigenstate $|a\rangle$. The probability of the prospect is composed of two factors: an utility factor, $f(\pi_a)$ (a factor containing the classical utility of a lottery) and an attraction factor, $q(\pi_a)$ (a probabilistic factor that results from the quantum interference effect). More specifically, for a lottery L_a, the utility factor $f(\pi_a)$ corresponds to minimizing the Kullback-Leibler information functional, which in the simple case of uncertainty yields (Yukalov and Sornette 2015): $f(\pi_a) = \frac{U(L_a)}{\sum_a U(L_a)}$. The attraction factor, on the other hand, represents the behavioural biases, which are expressed through quantum interference, and is a value between $-1 < q(\pi_a) < 1$. The final probability of the prospect is then given by $Pr(\pi_a) = f(\pi_a) + q(\pi_a)$.

In this work, considering $Pr(\pi_a)$ the classical probability distribution of the factor $\mu(X_2, A)$ and $f(\pi_a)$ the classical utility corresponding to the choice of some action A of the same factor $\mu(X_2, A)$, then we obtain

$$Pr(\pi_a) = Pr(X_1 = t) Pr(X_2|X_1 = t) + Pr(X_1 = f) Pr(X_2|X_1 = f)$$

$$f(\pi_a) = U(X_1 = t, A) + U(X_1 = f, A)$$

We can get the attraction factor, $q(\pi_a)$, by replacing the classical real numbers in $Pr(\pi_a)$ by quantum-like amplitudes. The quantum interference effects emerge by applying Born's rule,

$$q(\pi_a) = |\psi(X_1 = t)\psi(X_2|X_1 = t) + \psi(X_1 = f)\psi(X_2|X_1 = f)|^2$$

$$q(\pi_a) = |\psi(X_1 = t)\psi(X_2|X_1 = t)|^2 + |\psi(X_1 = f)\psi(X_2|X_1 = f)|^2 + 2Interf, \tag{17}$$

where the quantum interference term is given by

$$Interf = |\psi(X_1 = t)\psi(X_2|X_1 = t)| \, |\psi(X_1 = f)\psi(X_2|X_1 = f)| \cos(\theta_1 - \theta_2). \tag{18}$$

Since the factor $\mu(X_2, A)$ represented a probability distribution over the utility functions, we need to update the utility factor $f(\pi_a)$ in order to also represent this distribution over the quantum interference term. The quantum interference term, for N random variables grows (Moreira and Wichert 2016)

$$\sum_{i=1}^{N-1} \sum_{j=i+1}^{N} 2 \, |\psi(X_i = t)\psi(X_j|X_i = t)| \, |\psi(X_i = f)\psi(X_j|X_i = f)| \cos(\theta_i - \theta_j).$$

The utility factor $\mu(X_2, A)$ already specifies the utility function expressed in terms of the distribution of X_2. When we consider X_2 a quantum random variable, then this probability distribution is extended to incorporate the quantum interference effects. So, given that the utility function is expressed in terms of this probability distribution, we propose to update it in the same way as the interference term as

$$\sum_{i=1}^{N} U\left(X_i = t, A\right) U\left(X_i = f, A\right).$$

For our example, where we have utility factor $\mu\left(X_2, A\right)$ expressed in terms of the distribution of X_2, then

$$f(\pi_a) = U\left(X_1 = t, A\right) + U\left(X_1 = f, A\right) + U\left(X_1 = t, A\right) U\left(X_1 = f, A\right).$$

Under this representation, the result of the marginalisation, $\mu\left(X_2, A\right)$, will be given by the product of the vector representation of the utility factor $f(\pi_a)$ with the attraction factor $q(\pi_a)$:

$$\mu\left(X_2, A\right) = \langle q(\pi_a) | f(\pi_a) \rangle,$$

where the vector representation corresponds to

$$|q(\pi_a)\rangle = \begin{bmatrix} |\psi(X_1 = t)\psi(X_2|X_1 = t)|^2 \\ |\psi(X_1 = f)\psi(X_2|X_1 = f)|^2 \\ Interf \end{bmatrix} \qquad |f(\pi_a)\rangle = \begin{bmatrix} U\left(X_1 = t, A\right) \\ U\left(X_1 = f, A\right) \\ U\left(X_i = t, A\right) U\left(X_i = f, A\right) \end{bmatrix}.$$

This way, the final marginalisation for the quantum-like influence diagram is

$$\mu\left(X_2, A\right) = \langle q(\pi_a) | f(\pi_a) \rangle = |\psi(X_1 = t)\psi(X_2|X_1 = t)|^2 \cdot U\left(X_1 = t, A\right) + \\ |\psi(X_1 = f)\psi(X_2|X_1 = f)|^2 \cdot U\left(X_1 = f, A\right) + Interf \tag{19}$$

where

$$Interf = 2\,|\psi(X_1 = t)\psi(X_2|X_1 = t)|\,|\psi(X_1 = f)\psi(X_2|X_1 = f)|\,cos\left(\theta_1 - \theta_2\right).$$

Note that, in Eq. 20, if one sets the interference term $(\theta_i - \theta_j)$ to $\pi/2$, then $cos(\theta_i - \theta_j) = 0$. This means that the quantum interference term is canceled and the quantum-like influence diagram collapses to its classical counterpart.

Finally, the goal is to find the decision rule δ_A that maximizes $\mu\left(X_2, A\right)$,

$$\delta_A^*\left(a, Z\right) = \alpha(x) = \begin{cases} 1 & a = argmax\ \mu\left(X_2, A\right) \\ 0 & otherwise \end{cases} \tag{20}$$

We will refer to Eq. 20 as the quantum-like influence Equation, since we cannot call it a maximization of the expected utility in the real sense, because we had to change the distribution of the utility factor $\mu\left(X_2, A\right)$ in order to mach the distribution over the quantum interference terms.

7   A Quantum-Like Influence Diagram for the Prisoner's Dilemma Game

Several paradoxical findings have been reported over the literature showing that individuals do not act rationally in decision scenarios under uncertainty (Kuhberger et al. 2001; Tversky and Shafir 1992; Lambdin and Burdsal 2007; Hristova and Grinberg 2008; Busemeyer et al. 2006a). The quantum-like influence diagram can help to accommodate the several paradoxical decisions by manipulating the quantum interference effects that emerge from the inferences in the quantum-like Bayesian network, which can be used to reestimate the expected utilities.

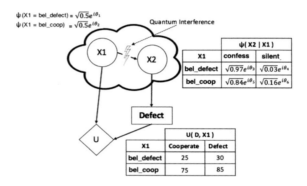

Fig. 3. Quantum-Like Influence Diagram representing the Prisoner's DIlemma Experiment from (Shafir and Tversky 1992). Random variable X_1 corresponds to the player's own believes about the action of his opponent (either believes the opponent will defect or the opponent will cooperate) and X_2 corresponds to the player's own strategy, i.e. either he confesses the crime or he remains silent.

We will model the works reported in Table 1 under the proposed quantum-like influence diagram. Figure 3 corresponds to the representation of the work of Shafir and Tversky (1992). The three types of nodes in the represented quantum-like influence diagram are the following:

- **Random Variables:** the circle-shaped nodes are the random variables belonging to the quantum-like Bayesian Network representing the player that needs to make a decision in the Prisoner's Dilemma, without being aware of the decision of his opponent. We modelled this network with two binary random variables, X_1 and X_2. X_1 corresponds to the player's own believes about the action of his opponent (either believes the opponent will defect or the opponent will cooperate) and X_2 corresponds to the player's own strategy, i.e. either he confesses the crime (and therefore would find it safe to engage in a *defect* strategy) or he remains silent (and would prefer to engage in a *cooperate* strategy). The tables next to each random variable are conditional probability tables and they show the probability distribution of the variable towards its parent nodes. These conditional probability tables match the probability distributions reported in Table 1. In the specific case of Fig. 3, this table is filled with the values of the probability amplitudes identified in the work of Shafir and Tversky (1992).

- **Action Node:** the rectangle shaped node is the action that we want to make a decision. In the context of the prisoner's dilemma we are interested to compute the maximum expected utility of defecting or not defecting (i.e. cooperating).

- **Utility Node:** the diamond shaped node corresponds to the payoffs that the player will have for taking (or not) the action $defect$, given his own personal strategy. The values in this node will be populated with the different payoffs used across the different experiments of the prisoner's dilemma game reported over the literature.

In the conditions where the player *knows* the strategy of his opponent, the quantum-like influence diagram collapses to its classical counterpart, since there is no uncertainty. This was already noticed in the previous works of Moreira and Wichert (2014, 2016, 2017). However, when the player is not informed about his opponent's decision, then the quantum-like Bayesian network will produce interference effects (Eq. 5). When computing the maximum expected utility, we will marginalise out X_1 like it was shown in Eq. 16. This will result in a factor showing the distribution of the player's personal strategy (either confess or remain silent) towards his believes over his opponents actions (either to defect or cooperate). The quantum interference term will play an important role to determine which quantum parameters can influence the player's decision to switch from a classical (and rational) defect action towards the paradoxical decision found in the works the literature, i.e. to cooperate.

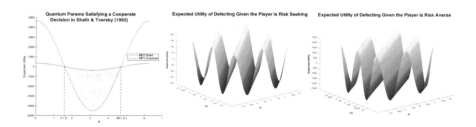

Fig. 4. Impact of quantum interference terms in the overall expected utility: (left) quantum parameters that maximize a cooperate decision, (center) variation of the expected utility when the player confesses (defects) and (right) variation of the expected utility when the player remains silent (cooperates).

Figure 4 demonstrates the impact of the quantum interference effects in the player's decision. The graphs in the centre and in the right of Fig. 4 represent all possible maximum expected utilities that the player can achieve by varying the quantum interference term θ in Eq. 20 for a personal strategy of confessing (defecting) or remaining silent (cooperating), respectively. On the left of Fig. 4, it is represented all the values of θ that satisfy the condition that $EU[Cooperate] > EU[Defect]$, i.e., all the values of the quantum interference parameter θ that will maximise the utility of cooperation rather than defect. One can note that, for experiment of Shafir and Tversky (1992) (as well as in the remaining works of the literature analysed in this work), one can maximise the expected utility of Cooperation when the utilities are negative. This is in accordance with the previous study of Moreira and Wichert (2016) in which the authors found that violations to the Sure Thing Principle imply destructive (or negative) quantum interference effects. As we will see in the next section, the quantum parameters found that are used to maximise the expected utility of a cooperate action lead to destructive quantum interferences and can exactly explain the probability distributions observed in the experiments.

7.1   Results and Discussion

Although there are several quantum parameters that satisfy the relationship that shows that participants can maximise the utility of a cooperate action, only a few parameters are able to accommodate both the paradoxical probability distributions reported in the several works in the literature and to maximise the expected utility of cooperating. The previous work of Moreira and Wichert (2016) shows how the quantum parameters are sensitive to accommodate the violations of the Sure Thing Principle in terms of the probability distributions. The slight variation of the quantum parameter θ in the quantum-like Bayesian network can lead to completely different probability distributions which differ from the ones observed in the difference experimental scenarios reported in the literature. These probability distributions will influence the utilities computed by the expected utility framework.

In Table 2, it is presented the quantum parameters that lead to the quantum interference term that is necessary to fully explain and accommodate the violations to the Sure Thing Principle reported over several works of the literature.

For this reason, we decided to test if the quantum-like parameters used to accommodate the violations to the Sure Thing Principle were sufficient and if they could also lead to a maximisation of expected utility of cooperation. We performed simulations of the different works in the literature and we concluded that the quantum interference effects that can accommodate violations to the violations of the Sure Thing Principle in the quantum-like Bayesian network alone, also explain a higher preference of the cooperative action over defect. Table 3 presents the results.

In Table 3, we present the Maximum Expected Utility (MEU) computed for each work in the literature using the classical approach for the player's different strategies: either remain silent (*CL silent*) or confess the crime (*CL confess*). The classical MEU shows that the optimal strategy is to *confess* and engage on a *defect* strategy independently of what action the opponent chose. Of course

Table 2. Experimental results reported for the Prisoner's Dilemma game. The entries highlighted correspond to games that are not violating the Sure Thing Principle.

	Prob of defect (Known to defect)	Prob of coop-erate (Known to cooperate)	Classical prob (Unknown condition)	Experim prob (Unknown condition)	Quantum interference θ param
Shafir and Tversky (1992)	0.9700	0.8400	0.9050	0.6300	2.8151
Li and Taplin (2002) G1	0.7333	0.6670	0.7000	0.6000	3.0170
Li and Taplin (2002) G2	0.8000	0.7667	0.7833	0.6300	3.0758
Li and Taplin (2002) G3	**0.9000**	**0.8667**	**0.8834**	**0.8667**	**2.8052**
Li and Taplin (2002) G4	0.8333	0.8000	0.8167	0.7000	3.2313
Li and Taplin (2002) G5	0.8333	0.7333	0.7833	0.7000	2.8519
Li and Taplin (2002) G6	**0.7667**	**0.8333**	**0.8000**	**0.8000**	**1.5708**
Li and Taplin (2002) G7	0.8667	0.7333	0.8000	0.7667	3.7812

these results go against the experimental works of the literature which say that a significant percentage of individuals, when under uncertainty, the engage more on cooperative strategies.

In opposition, when we use the quantum-like influence diagram, we take advantage of the quantum interference terms that will disturb the probabilistic outcomes of the quantum-like Bayesian networks. Since the utility function depends on the outcomes of the quantum-like Bayesian network, then it is straightforward that quantum interference effects influence indirectly the outcomes of the MEU allowing us to favour a different strategy predicted by the classical MEU.

Table 3. Inferences in the quantum-like influence diagram for different works of the literature reporting violations of the Sure Thing Principle in the Prisoner's Dilemma Game. One can see that the Quantum-Like Influence, presented in Eq. 20 (QL Infl) was changed to favour a Cooperate strategy using the quantum interference effects of the Quantum-Like Bayesian Network. In the payoffs, d corresponds to $defect$ and c to cooperate. The first payoff corresponds to player 1 and the second to player 2.

	Shafir and Tversky (1992)		Game 1		Game 2		Game 3		Game 4		Game 5		Game 6		Game 7	
	QL Infl (coop)	QL Infl (def)	QL Infl (coop)	QL Infl (def)	QL Infl (coop)	QL Infl (def)	QL Infl (coop)	QL Infl (def)	QL Infl (coop)	QL Infl (def)	QL Infl (coop)	QL Infl (def)	QL Infl MEU	QL Infl MEU	QL Infl (coop)	QL Infl (def)
CL confess	43.63	**50.25**	34.19	**39.35**	38.75	**61.78**	26.85	**50.33**	65.70	**67.33**	16.27	**34.50**	17.58	**36.50**	16.43	**35.00**
CL silent	6.38	**7.25**	15.82	**18.15**	11.25	**17.22**	3.65	**26.85**	14.80	**15.17**	5.23	**10.5**	3.92	**8.50**	5.07	**10.00**
QL confess	-1559.46	-2129.94	-1263.63	-1730.21	-1422.69	-4787.28	-702.24	-2075.58	-5198.14	-5462.41	-221.05	-1313.94	**28.83**	**36.49**	-184.75	-1116.33
QL silent	**116.66**	-160.08	-538.62	-735.89	-392.89	-1320.22	-94.44	-270.75	-1162.55	-1221.47	-61.44	-353.22	**3.91**	**8.50**	-44.86	-262.30
QL Interf $\theta_1 - \theta_2$	2.815	2.815	3.017	3.017	3.0758	3.0758	2.805	2.805	3.23	3.23	2.8519	2.8519	1.5708	1.5708	3.78	3.78
Payoff dd dc	30	25	30	25	73	25	30	25	80	78	43	10	30	10	30	10
cd cc	85	75	85	75	85	75	85	36	85	83	85	46	60	33	60	33

It is interesting to notice that indeed the parameters used accommodate the violations of the Sure Thing Principle alone in the quantum-like Bayesian Network could also be used to maximise the utility of a Cooperate action. This was verified in all works of the literature analysed except in Game 6 in the work of Li and Taplin (2002). The reason is that Game 6 is not even reporting a violation to the Sure Thing Principle and could be explained by the classical theory with a minor error percentage. This means that in this experiment, a *defect* action was favoured over a *cooperate* one.

8   Conclusion

In this work, we proposed an extension of the quantum-like Bayesian Network initially proposed by Moreira and Wichert Moreira and Wichert (2014, 2016) into a quantum-like influence diagram. Influence diagrams are designed for knowledge representation. They are a directed acyclic compact graph structure that represents a full probabilistic description of a decision problem by using probabilistic inferences performed in Bayesian networks (Koller and Friedman 2009) together with a fully deterministic utility function. Currently, influence diagrams have a vast amount of applications. They can be used to determine the value of imperfect information on both carcinogenic activity and human exposure (Howard and Matheson 2005), the are used to detect imperfections in manufacturing and they can even be used for team decision analysis (Detwarasiti and Shachter 2005), valuing real options (Lander and Shenoy 2001), etc.

The preliminary results obtained in this study show that the quantum-like Bayesian network can be extended to deal with decision-making scenarios by incorporating the notion of maximum expected utility in influence diagrams. The general idea is to take advantage of the quantum interference terms produced in the quantum-like Bayesian network to influence the probabilities used to compute the expected utility. This way, we are not proposing a new type of expected utility hypothesis. On the contrary, we are keeping it under its classical definition. We are only incorporating it as an extension of a quantum-like probabilistic graphical model where the utility node depends only on the probabilistic inferences of the quantum-like Bayesian network.

This notion of influence diagrams opens several new research paths. One can incorporate different utility nodes being influenced by different random variables of the quantum-like Bayesian Network. This way one can even explore different interference terms affecting different utility nodes, etc. We plan to carry on with this study and further develop these ideas in future research.

References

Bruza, P., Wang, Z., Busemeyer, J.: Quantum cognition: a new theoretical approach to psychology. Trends Cogn. Sci **19**, 383–393 (2015)

Busemeyer, J., Bruza, P.: Quantum Model of Cognition and Decision. Cambridge University Press, Cambridge (2012)

Busemeyer, J., Matthew, M., Wang, Z.: A quantum information processing explanation of disjunction effects. In: Proceedings of the 28th Annual COnference of the Cognitive Science Society (2006a)

Busemeyer, J., Wang, Z., Lambert-Mogiliansky, A.: Empirical comparison of Markov and quantum models of decision making. J. Math. Psychol. **53**, 423–433 (2009)

Busemeyer, J., Wang, Z., Townsend, J.: Quantum dynamics of human decision making. J. Math. Psychol. **50**, 220–241 (2006b)

Detwarasiti, A., Shachter, R.D.: Influence diagrams for team decision analysis. Decis. Anal. **2**, 207–228 (2005)

Deutsch, D.: Quantum theory of probability and decisions. Proc. Roy. Soc. A **455**, 3129–3137 (1988)

Howard, R., Matheson, J.: Influence diagrams. In: Readings on the Principles and Applications of Decision Analysis. Strategic Decisions Group (1984)

Howard, R., Matheson, J.: Influence diagrams. Decis. Anal. **2**, 127–143 (2005)

Hristova, E., Grinberg, M.: Disjunction effect in prisoner's dilemma: evidences from an eye-tracking study. In: Proceedings of the 30th Annual Conference of the Cognitive Science Society (2008)

Kahneman, D., Slovic, P., Tversky, A.: Judgment Under Uncertainty: Heuristics and Biases. Cambridge University Press, Cambridge (1982)

Kahneman, D., Tversky, A.: Prospect theory - an analysis of decision under risk. Econometrica **47**, 263–292 (1979)

Koller, D., Friedman, N.: Probabilistic Graphical Models: Principles and Techniques. The MIT Press, Cambridge (2009)

Kuhberger, A., Komunska, D., Josef, P.: The disjunction effect: does it exist for two-step gambles? Organ. Behav. Hum. Decis. Process. **85**, 250–264 (2001)

Lambdin, C., Burdsal, C.: The disjunction effect reexamined: relevant methodological issues and the fallacy of unspecified percentage comparisons. Organ. Behav. Hum. Decis. Process. **103**, 268–276 (2007)

Lander, D., Shenoy, P.: Modeling and Valuing Real Options-Using Influence Diagrams. Technical report, School of Business, Babson College (2001)

Li, S., Taplin, J.: Examining whether there is a disjunction effect in prisoner's dilemma game. Chin. J. Psychol. **44**, 25–46 (2002)

Moreira, C., Wichert, A.: Interference effects in quantum belief networks. Appl. Soft Comput. **25**, 64–85 (2014)

Moreira, C., Wichert, A.: Quantum-like Bayesian networks for modeling decision making. Front. Psychol. **7**, 11 (2016)

Moreira, C., Wichert, A.: Exploring the relations between quantum-like Bayesian Networks and decision-making tasks with regard to face stimuli. J. Math. Psychol. **78**, 86–95 (2017)

Mura, P.L.: Projective expected utility. J. Math. Psychol. **53**, 408–414 (2009)

Pothos, E., Busemeyer, J.: A quantum probability explanation for violations of rational decision theory. Proc. Roy. Soc. B **276**, 2171–2178 (2009)

Savage, L.: The Foundations of Statistics. Wiley, Hoboken (1954)

Shafir, E., Tversky, A.: Thinking through uncertainty: nonconsequential reasoning and choice. Cogn. Psychol. **24**, 449–474 (1992)

Tversky, A., Kahneman, D.: Judgment under uncertainty: heuristics and biases. Science **185**, 1124–1131 (1974)

Tversky, A., Shafir, E.: The disjunction effect in choice under uncertainty. J. Psychol. Sci. **3**, 305–309 (1992)

von Neumann, J., Morgenstern, O.: Theory of Games and Economic Behavior. Princeton University Press, Princeton (1953)

Yukalov, V., Sornette, D.: Decision theory with prospect interference and entanglement. Theor. Decis. **70**, 283–328 (2011)

Yukalov, V., Sornette, D.: Preference reversal in quantum decision theory. Front. Pschol.: Cogn. **6**, 1538 (2015)

Zurek, W.: Entanglement symmetry, amplitudes, and probabilities: inverting Born's rule. Phys. Rev. Lett. **106**, 1–5 (2011)

Cybernetics and AI

Fuzzy Logic Behavior
of Quantum-Controlled Braitenberg
Vehicle Agents

Rebeca Araripe Furtado Cunha[1,2], Naman Sharma[1,3], Zeno Toffano[1,4(✉)],
and François Dubois[5,6]

[1] CentraleSupélec, Université ParisSaclay, 91190 Gif-sur-Yvette, France
zeno.toffano@centralesupelec.fr, zeno.toffano@supelec.fr
[2] Federal University of Rio de Janeiro, Rio de Janeiro 21941-909, Brazil
[3] National University of Singapore, Singapore 119077, Singapore
[4] Laboratoire des Signaux et Systèmes - CNRS (UMR8506), Gif-sur-Yvette, France
[5] Conservatoire National des Arts et Métiers, 75003 Paris, France
francois.dubois@u-psud.fr
[6] Association Française de Science des Systèmes (AFSCET), Paris, France

Abstract. The behavior of agents represented by Braitenberg vehicles
is investigated in the context of the quantum robot paradigm. The agents
are processed through quantum circuits with fuzzy inputs, this permits
to enlarge the behavioral possibilities and the associated decisions for
these simple vehicles. The logical formulation Eigenlogic, using quantum logical observables as propositions and eigenvalues as truth values is applied in this investigation. Fuzzy logic arises naturally in this
formulation when considering input states that are not eigenvectors of
the logical observables, the fuzzy membership being the quantum mean
value of the logical observable on the input state. Computer simulations
permits visualization of complex behaviors resulting from the multiple
combination of quantum control gates. This allows the detection of new
Braitenberg vehicle behavior patterns related to identified emotions and
linked to quantum-like effects.

Keywords: Quantum robots · Fuzzy logic · Quantum gates ·
Braitenberg vehicles · Emotion analysis

1 Introduction

With the recent improvements in quantum information, interest has been grow-
ing in the area of quantum robotics. This signifies using concepts from quantum computing to build and conceptualize systems capable of decision making.
Simple robots capable of showing complex behavior have been introduced in
Valentino Braitenberg's work on cybernetics proposing vehicle agents that show
human-like emotions [1]. Here we aim to implement *Braitenberg Vehicles* (BV)
using quantum-like circuits. Paul Benioff who was the first to propose the idea

© Springer Nature Switzerland AG 2019
B. Coecke and A. Lambert-Mogiliansky (Eds.): QI 2018, LNCS 11690, pp. 111–122, 2019.
https://doi.org/10.1007/978-3-030-35895-2_8

of a quantum Turing machine in 1980 was also the proponent of the theoretical principle of a *quantum robot* [2].

A quantum-like implementation of BVs has been undertaken in a previous research work [3], where Raghuvanshi et al. used the reversible quantum logical gate structure in building prototype BVs with Lego blocks. In the work presented here, in order to provide flexibility to the vehicle's behavior, we design fuzzy logic quantum-like circuits based on the quantum models proposed in [4].

Our goal here is to test the multiple combinations of quantum gates used in the control of BV by analyzing their complex behavior. For this purpose, we have developed a visual simulation tool illustrating BV's behaviors. This allows us to investigate new emotional behavior patterns not expected by a simple theoretical analysis used for the description of BV in [1].

Mathematical models and simulations of individual and swarm automaton agents in response to environmental stimuli have attracted much interest for the understanding of complex behaviors of a group of animals. In [5], Kangan et al. developed a probabilistic control model (denominated ANIMAT) of mobile agents with biologically-inspired navigation, with the goal to mimic an ant colony. In this research it was made clear that the agent changes its state as a response to environmental stimuli and/or as a result of its own action on the environment, this observation can lead to path planning and intelligent-like emergent behavior of a group of agents. So agent behavioral models should take into account the uncertainty and the dynamics due to their environment.

A theoretical description of the model proposed here for the interaction of BV quantum robots with their environment is given in Sect. 2. This section details the mechanisms of transformation from the sensor input stimuli states to the control signals for the wheels of the BV (see Fig. 2).

In Sect. 3 we provide a brief outline of the logical models used in the design of the BV quantum robots. These models are based on the *Eigenlogic* approach proposed in [4] and permit to characterize behaviors associated to fuzzy logic control.

We finish by illustrating the simulation results in Sect. 4. In order to test the accuracy of our simulations, we first verify the behaviors predicted by the thought experiments proposed by Valentino Braitenberg [1]. This is done by observing how our simulator reacts to logical operators specifically designed for representing the vehicle's emotions of Fear, Aggression, Love and Exploration. After testing the BVs with these well established emotions, we test other Eigenlogic control operators such as *i.e.* the implication operator (see Dubois and Toffano in [4]). Finally using different operators in a variety of combinations gives us an interesting insight into the global emotional behavior of the vehicles.

2   Modeling of the Vehicle

2.1   Introduction to Braitenberg Vehicles

In his book "Vehicles: Experiments in Synthetic Psychology" [1] Valentino Braitenberg describes various thought experiments using simple machines (*Braiten-*

berg Vehicles BV) that consist of sensors, motors and wheels. Similarly in our approach the quantum robot agents are represented by two controlled wheels at the rear of the BV, with two sensors at the front. The sensors detect light produced by surrounding sources. The sensors can be connected in different combinations to the wheels, and may have a positive or negative relation with the strength of the stimuli. These simple changes in configuration can lead to complex and surprising results in the agent behavior. Braitenberg terms this the "law of uphill analysis and downhill invention". According to this principle it is far easier to create machines that exhibit complex behavior based on simple connective structures than to try to derive their structures from behavioral observations and interpretations.

2.2 Quantum Approach for BV Decision Making

To model the system we consider BVs with two sensors: SL in the upper left corner and SR in the upper right corner (see Fig. 2). The sensors detect light intensity and transform it into an input state vector for the quantum circuit. The system then delivers as an output a fuzzy logical measure that is translated into wheel control by motors.

The computational block is composed of matrix operators designed using the quantum-like Eigenlogic method [4] discussed in Sect. 3. We use two operators $\mathbf{F}_L$ and $\mathbf{F}_R$. Each one takes the inputs and delivers signals to the respective motors (left or right). This approach avoids to physically permute the wiring connecting sensors and motors in order to obtain different behaviors, it all can be handled by the computing device. The diagram in Fig. 1 resumes the input-output process.

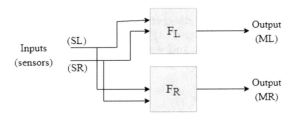

Fig. 1. Simplified diagram of the control circuit of a BV connected to the left and right input sensors SL, SR and to the output left and right wheel motors ML, MR.

The equivalent quantum formulation of the system consists in defining the input state vector presented in Eq. (1),

$$|x\rangle = |x_L\rangle \otimes |x_R\rangle = |x_L x_R\rangle \tag{1}$$

where the *ket* $|x\rangle$ corresponds to the combined input state of the bilinear system, formed by the Kronecker product of the individual input states $|x_L\rangle$ and $|x_R\rangle$

corresponding respectively to the left and right sensors. These vectors can be understood as qubits with a given orientation in Hilbert space. The output signal is obtained by the quantum mean value (Born rule) of the logical projection observable on the compound input state $|x\rangle$. It is the control function for the wheels. The quantities μ_L and μ_R, for left and right wheel control, are:

$$\mu_L = \langle x| \mathbf{F}_L |x\rangle \tag{2}$$

$$\mu_R = \langle x| \mathbf{F}_R |x\rangle \tag{3}$$

These quantities can also be interpreted as *fuzzy membership functions* [4]. In the following section we will show that these values are proportional to the angular speed of the wheels.

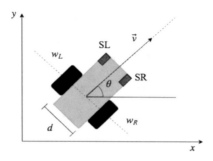

Fig. 2. Braitenebrg Vehicle orientation at speed $\vec{v}$ showing at the rear the left and right wheels WL, WR and at the front the light sensors SL, SR spaced at distance d.

2.3 Output States and Vehicle's Motion

Once the vehicle is submitted to the input state $|x_f\rangle$, it feeds it into the computation block as presented in the diagram of Fig. 1. The mean values for left and right are obtained as in Eqs. (2) and (3), depending directly in the logical control operators $\mathbf{F}_L$ and $\mathbf{F}_R$. The wheel angular speeds ω_L and ω_R are calculated by applying the following relations:

$$\omega_L = \mathrm{K} \cdot \mu_L \tag{4}$$

$$\omega_R = \mathrm{K} \cdot \mu_R \tag{5}$$

where K is a constant which allows us to tune the sensitivity of the vehicle to the stimuli. The x and y components of the vehicle speed vector (cf. Fig. 2) are determined by the following differential equations:

$$\dot{x} = v_x = \frac{1}{2} R \left(\omega_L + \omega_R\right) cos(\theta) \tag{6}$$

$$\dot{y} = v_y = \frac{1}{2} R \left(\omega_L + \omega_R\right) sin(\theta) \tag{7}$$

$$\dot{\theta} = \frac{1}{d} R \left(\omega_R - \omega_L\right) \tag{8}$$

where R represents the BV wheel radius.

The period of time between every stimuli update is considered the unit of time of our system. Thus the vehicle position frame update equations are given by:

$$x(t_{n+1}) = x(t_n) + v_x(t_n) \cdot (t_{n+1} - tn) = x(t_n) + v_x(t_n) \qquad (9)$$

$$y(t_{n+1}) = y(t_n) + v_y(t_n) \cdot (t_{n+1} - tn) = y(t_n) + v_y(t_n) \qquad (10)$$

3   Topics in Eigenlogic

Eigenlogic, [4] is a new quantum-like formulation in logic associating linear algebra and propositional logic. In this picture logical connectives are represented by observables and the truth values of the propositions are the eigenvalues of these observables. This permits to make a straightforward correspondence with quantum observables. As explained previously in Sect. 2, the computational block of our simulated vehicles is composed of two-argument logical observables.

3.1   The Eigenlogic Operators

Because of logical completeness an n-argument (arity-n) binary logical system has a total of 2^{2^n} logical connectives. In Eigenlogic this is related to the number of possible compatible projection operators, which are commuting observables. For example, for a two qubit system defined in a 4-dimensional space, the number of possible logical connectives is $2^{2^2} = 16$. In this case, we have to consider $2^2 = 4$ different distinct input cases corresponding to the 4 eigenstates of the logical system: $|00\rangle$, $|01\rangle$, $|10\rangle$ and $|11\rangle$. For each propositional case (eigenstate) we can assign a truth value which corresponds to one of the eigenvalues $\{0, 1\}$.

For a given logical connective, we derive the corresponding logical observable by matrix interpolation methods using the eigenvalues and the associated projection operators as discussed in [6]. In the case of two logical input arguments A and B, the basic logical projector operators and its negations are given in Eq. (11), where $\mathbb{I}$ is the identity operator.

$$\begin{aligned}
\mathbf{F}_A &= |1\rangle \langle 1| \otimes \mathbb{I} & \mathbf{F}_{\bar{A}} &= (\mathbb{I} - |1\rangle \langle 1|) \otimes \mathbb{I} = |0\rangle \langle 0| \otimes \mathbb{I} \\
\mathbf{F}_B &= \mathbb{I} \otimes |1\rangle \langle 1| & \mathbf{F}_{\bar{B}} &= \mathbb{I} \otimes (\mathbb{I} - |1\rangle \langle 1|) = \mathbb{I} \otimes |0\rangle \langle 0|
\end{aligned} \qquad (11)$$

In general every logical connective can be expressed as a linear combination of the preceding observables using the conjunction connective. In logic this corresponds to the canonical form: disjunction of conjunctions SOP (Sum Of Products). In a 2-argument Eigenlogic system, one will consider the 4 conjunction observables corresponding to the rank-1 projection operators $\mathbf{F}_{A \cdot B} = \mathbf{F}_A \cdot \mathbf{F}_B$, $\mathbf{F}_{\bar{A} \cdot B}$, $\mathbf{F}_{A \cdot \bar{B}}$ and $\mathbf{F}_{\bar{A} \cdot \bar{B}}$. The scalar coefficients of the linear combination are the truth-values of the logical connective one wants to express. The linear interpolation used for the "implication" logical observable is given by:

$$\mathbf{F}_{A \implies B} = 1\,\mathbf{F}_{A \cdot B} + 1\,\mathbf{F}_{\bar{A} \cdot B} + 0\,\mathbf{F}_{A \cdot \bar{B}} + 1\,\mathbf{F}_{\bar{A} \cdot \bar{B}} = \mathbb{I} - \mathbf{F}_A + \mathbf{F}_{A \cdot B} \qquad (12)$$

3.2   Incorporation of Fuzzy Logic

In order to better represent real situations, we consider that the stimuli of the BV's light sensors correspond to continuous variations of the light intensity. In our quantum formulation the vector eigenspace of a 2-argument Eigenlogic observables is the 2-qubit canonical basis $\{|00\rangle, |01\rangle, |10\rangle, |11\rangle\}$. State $|0\rangle$, corresponds to no light at all and state $|1\rangle$, corresponds to maximum luminosity.

Fuzzy logic deals with truth values that may be any number between 0 and 1, where the truth of a proposition may range between completely false and completely true. In quantum mechanics one can always express a state-vector as a decomposition on an orthonormal basis. To make a correspondence with fuzzy characteristics we will consider not only eigenstates as input states but in general linear combinations of eigenstates, as shown in (13).

$$|x_f\rangle = \alpha_{00} |00\rangle + \alpha_{01} |01\rangle + \alpha_{10} |10\rangle + \alpha_{11} |11\rangle \tag{13}$$

The coefficients of the development can be considered as the weight of a particular logical state. The square module of each coefficient corresponds to the probability of being in that state and the sum of these probabilities will add up to one bacause of the othonormalization of the vector states. In our case, the compound fuzzy input state $|x_f\rangle$ is the Kronecker tensor product of the vectors $|x_L\rangle$ and $|x_L\rangle$ as defined in Eq. (1). So we can write the state vectors:

$$|x_L\rangle = \sqrt{1 - p_L} \, |0\rangle + \sqrt{p_L} \, |1\rangle = \begin{bmatrix} \sqrt{1 - p_L} \\ \sqrt{p_L} \end{bmatrix} \tag{14}$$

$$|x_R\rangle = \sqrt{1 - p_R} \, |0\rangle + \sqrt{p_R} \, |1\rangle = \begin{bmatrix} \sqrt{1 - p_R} \\ \sqrt{p_R} \end{bmatrix} \tag{15}$$

where the coefficients are function of the probabilities $p_L = |\langle x_L |1\rangle|^2$ and $p_R = |\langle x_R |1\rangle|^2$. The input state of the compound fuzzy system is then written as:

$$|x_f\rangle = |x_L x_R\rangle = \begin{bmatrix} (1 - p_L)(1 - p_R) \\ (1 - p_L)p_R \\ p_L(1 - p_R) \\ p_L p_R \end{bmatrix}^{\frac{1}{2}} \tag{16}$$

4   BV Quantum Robot Simulation Results

We performed the simulation with different configurations of logical observables in the vehicle's computational unit. In this paper we illustrate only a small subset of the rich variety of possible combinations of observables. The graphical simulation interface allows the user to easily select the desired control operator corresponding to each wheel, as well as to add light sources and new vehicles in any time slot and location on the canvas ot the running simulation.

4.1   Simulation Environment

As a result of the principle of the "Law of uphill analysis and downhill invention" [1], the behavior of these vehicles can become very complex. It is thus difficult to truly understand the response of these vehicles by the means of only thought experiments. Consequently, the natural method to analyze these vehicles is through computer simulations. The environment stimuli are light sources placed at different positions on the canvas. We opt for non punctual sources, where the intensity of each source is 1.0 inside the circle that delimits its border and decreases with the square of the distance to the border. The sources are considered as incoherent since they don't interact with each other, the total intensity at a given position is the sum of the intensities. Every vehicle has two sensors, one attached to its left and the other to its right side. The distance between both sensors is sufficient for the vehicle to distinguish the light intensities, in our model proportional to p_L and p_R, in each of the corners. Different intensities will result in a change of the angle of the vehicle's speed vector. To each considered behavior corresponds a characteristic truth table given by the logical operator mean values on the eigenstates.

We will now describe the basic behaviors associated to BVs.

4.2   Illustration of Different Emotional Behaviors

Fear: $F_L = F_A$ $F_R = F_B$. In this configuration, the quantum control gates simply connect the sensor readings to the wheel on the same side. Here, we see that the value of μ_R (resp. μ_L) corresponds to p_R (resp. p_L), because the logical control connective corresponds to the logical input. In Fig. 3, we clearly see the vehicle trying to avoid the stimulus in the case where the source is placed at a skew angle. In the case where the source is directly in front of the vehicle, the vehicle moves towards this source. The vehicle then comes to a rest in an area with almost no stimulus. This is the expected behavior for a BV possessing the emotion *Fear*.

Aggression: $F_L = F_B$ $F_R = F_A$. This configuration is similar to *Fear* with the difference that the quantum control gates connect the sensor output to the motor of the wheel on the opposite side. In Fig. 3, we observe a vehicle trying to collide with the source, regardless of the position of the source with respect to the vehicle. Also in this case the vehicle comes to a rest in an area with almost no stimulus. This is the expected behavior for BV possessing the emotion *Aggression*.

Passion: $F_L = F_B \nRightarrow A$ $F_R = F_A \nRightarrow B$. According to simulations, as illustrated in Fig. 3, this vehicle remains at rest in the absence of stimulus. When it detects a light source, it moves towards it with increasing speed. The vehicle eventually hits the source and then comes to a rest. Observing the behavior of this vehicle, we compared this BV to the emotion *Love* [1]: the vehicle goes towards the light source and stops when it reaches the source. But in the present

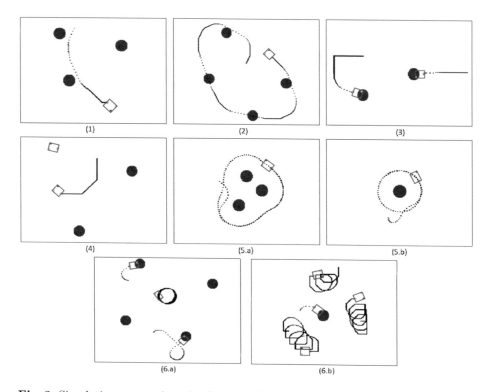

Fig. 3. Simulation screen-shots for (1) Fear, (2) Aggression, (3) Passion, (4) Boredom, (5) Worship, (6) Doubt

case it also shows aggressiveness because the speed increases as it gets closer to the source. Hence, we associate this BV emotion with *Passion*.

Boredom: $\mathbf{F}_L = \mathbf{F}_{A \nrightarrow B}$ $\mathbf{F}_R = \mathbf{F}_{B \nrightarrow A}$. In a thought experiment, one can expect that switching $(L \leftrightarrow R)$ the Eigenlogic operators of the *Passion* vehicle should lead to a result similar to that performed by the *Explore* vehicle described by Braitenberg [1], but with its own peculiarities. As one can see in Fig. 3, the vehicle remains at rest in the absence of light. In the presence of light it accelerates for a short period of time but then abruptly decelerates and comes to a stop before it actually hits the source. In the case another stronger source suddenly appears, the agent moves towards this source and then occupies a position at half-distance from both sources and facing away from the sources. Here, we interpret this BV emotion with *Boredom*.

Worship: $\mathbf{F}_L = \mathbf{F}_{HH}$ $\mathbf{F}_R = \mathbf{F}_B$. Here, the control logical operator is the normalized projector version of the double-Hadamard quantum gate $\mathbf{H} \otimes \mathbf{H}$:

$$\mathbf{F}_{HH} = \frac{1}{2}\left(\mathbb{I} - \mathbf{H} \otimes \mathbf{H}\right) \tag{17}$$

The inputs of the vehicle can be represented by the truth-table given in Table 1 proportional to the respective fuzzy measures on the control logical operators $\mu_L = \langle x| \mathbf{F}_{HH} |x\rangle$ and $\mu_R = \langle x| \mathbf{F}_B |x\rangle$:

Table 1. *Worship* emotion truth table

$|x_f\rangle$	μ_L	μ_R	Behavior
$|00\rangle$	0.25	0	Turns to the right slowly
$|01\rangle$	0.75	1	Turns to the left slowly
$|10\rangle$	0.75	0	Turns to the right
$|11\rangle$	0.25	1	Turns to the left

The vehicle keeps rotating around its own center in the absence of light. In the presence of light, it goes towards the source and starts to rotate around the source (or multiple sources when they are close together). This behavior can be seen in Fig. 3. We associate this vehicle emotion with *Worship* since the vehicles immediately start revolving around the light source once it senses it.

Doubt: $\mathbf{F}_L = \mathbf{F}_{B \not\Rightarrow A}$ $\mathbf{F}_R = \mathbf{F}_{XOR}$. The projective version of the exclusive-disjunction (XOR) self-inverse Eigenlogic operator $\mathbf{Z} \otimes \mathbf{Z}$ [4] is used here:

$$\mathbf{F}_{XOR} = \frac{1}{2} \left(\mathbb{I} - \mathbf{Z} \otimes \mathbf{Z} \right) \tag{18}$$

This operator provides a property that makes the vehicle turn around in circles, regardless of the presence or absence of stimuli. It is the other wheel control operator that completes the overall behavior of this vehicle. In the case considered here, we use the implication operator $\mathbf{F}_{B \not\Rightarrow A}$ as control operator for the left wheel. The truth table for this vehicle is given in Table 2.

This agent keeps rotating around itself in the absence of light. However, once it senses stimuli, it starts to roll towards it. Once it gets close to the source, it rotates outwards again and goes away from the source in a circular orbit. In some cases, the agent hits the source and comes to a stop (Fig. 3(6.a)). In particular cases, it starts rotating in a circle whose center itself is moving in a circle around the stimuli (Fig. 3(6.b)). We associate this vehicle with the emotional feeling of *Doubt*. The BV is unable to decide if it prefers light or not, and hence keeps coming to and going away from sources of light. In certain conditions, it decides to orbit around the source. In the case of sources on its both sides, it is unable to decide between the two and keeps switching from one to the other.

Table 2. *Doubt* emotion truth table

$|x_f\rangle$	μ_L	μ_R	Behavior
$|00\rangle$	0	0	No movement
$|01\rangle$	1	1	Goes straight
$|10\rangle$	0	1	Turns to the left
$|11\rangle$	0	0	No movement

5 Quantum Wheel of Emotions

The concept of "wheel of emotions" introduced by Plutchik et al. [7] pictures the idea that a complex emotional state is the composition of elementary emotions. This picture can be interpreted in a quantum-like way using the quantum state vector $|\psi\rangle$. Each qubit (2-dimensional) quantum state can be mapped to a point on the surface of the Bloch unit sphere:

$$|\psi\rangle = cos(\frac{\theta}{2})\,|0\rangle + e^{-i\frac{\phi}{2}}\,sin(\frac{\theta}{2})\,|1\rangle \tag{19}$$

where ϕ and θ are the spherical angles. In order to simplify interpretation, the coefficients (associated with the degree of truth) multiplying the base states are taken as real numbers. The points of the vector are thus placed on a circle corresponding to a *quantum wheel of emotions*.

We summarized our simulation results by associating the different observed behaviors to a sector in the wheel as shown in Fig. 4. Other emotions not presented in Fig. 3 have been simulated such as for example: *Interest, Curiosity, Distraction, Fear, Worship* and *Sadness*. The two latter ones have been obtained using a circuit that combines the standard quantum gates **H** and $\mathbf{C}_{\mathrm{NOT}}$.

The quantum wheel of emotions thus allows a continuous set of emotional states. A small perturbation in the angle of the input state $|\psi\rangle$ due to environmental factors, even if still inside the same emotional sector, will correspond to small changes in the vehicle's behavior. The measurement of the input state implies the collapse of $|\psi\rangle$ to a specific point of the wheel, and thus we can say that, in this aspect, the vehicle behaves as a quantum-like system. Furthermore, the fuzzy aspect of the system arouses naturally since the collapse can involve any state belonging to the continuous surface of the wheel. These observations can be compared to the observed similarities between neural network models and quantum systems. In particular, it has been suggested that it is possible to implement quantum learning algorithms dedicated to fuzzy qubits [8] where the weighted sums of inputs of a neuron correspond to the superposition of quantum states at the input of a quantum circuit and the quantum wave function collapse corresponds to the threshold activation of a neuron.

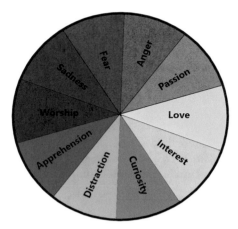

Fig. 4. Quantum Wheel of Emotions. For a given emotion, we associate quantum operators controlling the speed of the left and the right wheel respectively. **(Anger)** F_B, F_A; **(Passion)** $F_B \nRightarrow A$, $F_A \nRightarrow B$; **(Love)** $F_{\bar{A}}$, $F_{\bar{B}}$; **(Interest)** $F_{\bar{B}}$, $F_{\bar{A}}$; **(Curiosity)** $F_A \nRightarrow B$, $F_B \nRightarrow A$; **(Distraction)** $F_A \nRightarrow B$, F_A XOR B; **(Apprehension)** $F_B \nRightarrow A$, F_A XOR B; **(Worship)** $H \otimes H$, F_B; **(Sadness)** C_{NOT}, C_{NOT}; **(Fear)** F_A, F_B.

6   Discussion and Conclusion

The purpose of this research is to show the multiplicity of behaviors obtained by using fuzzy logic along with quantum logical gates in the control of simple *Braitenberg Vehicle* agents. The number of cases becomes intractable in simple theoretical approaches with increasing complexity. A computer simulation is mandatory and allows us to abstract the complexity by observing the motion of the vehicles and use it for illustrative purposes. At the same time, we see that by changing and combining different quantum control gates we can tune small changes in the vehicle's behavior, and hence get specific features around the main basic BV emotions of *Fear, Aggression, Love* and *Explore*. By tweaking these quantum gates, one can also obtain a vehicle that has a mixture of multiple emotions.

Further extensions to this project can be imagined. Currently, when the vehicles collide, their respective control operators could change in order to reflect a quantum-like entanglement behavior due to interaction. It would be interesting to entangle the vehicles so that the behavior of one vehicle depends upon the current state of the environment from the perspective of other vehicles even after they separate after the collision. It could also be interesting to explore the Braitenberg vehicles using different types of stimuli (instead of only light) and sensors. Also a formalization of the quantum BV components as quantum neural networks could lead to new investigation strategies and could benefit researches in machine learning algorithms related to emotion analysis.

7   Credits and Acknowledgements

The first coauthor is under the Brafitec scholarship, CAPES Foundation, Ministry of Education of Brazil, Brasilia 70.040-020, Brazil.

We want to thank the AFSCET (Association Française de Science des Systèmes) for permitting us to present the idea of this work at the WOSC (World Organisation of Systems and Cybernetics) 2017 Congress in Rome Italy.

We are grateful to Francesco Galofaro of LUB for fruitful discussions on logic and semantics and their link to quantum theory and for pointing out that the late Professor Valentin von Braitenberg was one of the academic founding members of LUB (Libera Università di Bolzano/Freie Universität Bozen, Italy).

References

1. Braitenberg, V.: Vehicles - Experiments in Synthetic Psychology. MIT Press, Cambridge (1986)
2. Benioff, P.: Quantum robots and environments as a first step towards a quantum mechanical description of systems that are aware of their environment and make decisions. Phys. Rev. A **58**(2), 893–904 (1998)
3. Raghuvanshi, A., Fan, Y., Woyke, M., Perkowski, M.: Quantum robots for teenagers. In: 37th International Symposium on Multiple-Valued Logic ISMVL (2007)
4. Dubois, F., Toffano, Z.: Eigenlogic: a quantum view for multiple-valued and fuzzy systems. In: de Barros, J.A., Coecke, B., Pothos, E. (eds.) QI 2016. LNCS, vol. 10106, pp. 239–251. Springer, Cham (2017). https://doi.org/10.1007/978-3-319-52289-0_19
5. Kagan, E., Rybalov, A., Sela, A., Siegelmann, H.: Probabilistic control and swarm dynamics in mobile robots and ants. In: Biologically-Inspired Techniques for Knowledge Discovery and Data Mining, January, pp. 11–13 (2014)
6. Toffano, Z., Dubois, F.: Interpolating binary and multivalued logical quantum gates. In: Proceedings of the 4th International Electronic Conference on Entropy and Its Applications. MDPI Proceedings, vol. 2, no. 4, p. 152 (2018)
7. Plutchik, R.: The nature of emotions. Am. Sci. **89**(4), 334–350 (2001)
8. Hannachi, M.S., Dong, F., Hirota, K.: Emulating quantum interference and quantum associative memory using fuzzy qubits. In: Proceedings of the IEEE International Conference on Computational Cybernetics ICCC 2007, pp. 39–45 (2007)

Moral Dilemmas for Artificial Intelligence: A Position Paper on an Application of Compositional Quantum Cognition

Camilo M. Signorelli[1,2,5(✉)] 🆔 and Xerxes D. Arsiwalla[3,4,5]

[1] Department of Computer Science, University of Oxford, Oxford, UK
camiguel@uc.cl
[2] Cognitive Neuroimaging Lab, INSERM U992, NeuroSpin,
Gif-sur-Yvette, France
[3] Barcelona Institute of Science and Technology, Barcelona, Spain
[4] Institute for Bioengineering of Catalonia, Barcelona, Spain
[5] Pompeu Fabra University, Barcelona, Spain

Abstract. Traditionally, the way one evaluates the performance of an Artificial Intelligence (AI) system is via a comparison to human performance in specific tasks, treating humans as a reference for high-level cognition. However, these comparisons leave out important features of human intelligence: the capability to transfer knowledge and take complex decisions based on emotional and rational reasoning. These decisions are influenced by current inferences as well as prior experiences, making the decision process strongly subjective and "apparently" biased. In this context, a definition of compositional intelligence is necessary to incorporate these features in future AI tests. Here, a concrete implementation of this will be suggested, using recent developments in quantum cognition, natural language and compositional meaning of sentences, thanks to categorical compositional models of meaning.

Keywords: Moral dilemmas · Moral test · Turing test · Artificial Intelligence · Compositional semantics · Natural language · Quantum cognition

1 Introduction

Moral dilemmas and a general intelligence definition have been recently suggested as an alternative to current AI tests [1]. Usually, Intelligence is interpreted regarding particular and efficient behaviours which can be measured in terms of performing or not these behaviours. One example is the Turing test [2], in fact, the first approach grounded on human behaviour and the most well-known and controversial test for AI. Other examples are challenging machines in games like chess or Go [3], and testing AI programs with dilemmas [4], theoretically, demanding a more complex level of information processing. Nevertheless, all these approaches lack a general definition of intelligence as a minimal requirement to measure intelligence [5–7], and restrict AI only to humankind "intelligence" without including key features of a true human intelligence. Therefore, this position paper is going to shortly introduce a new strategy

© Springer Nature Switzerland AG 2019
B. Coecke and A. Lambert-Mogiliansky (Eds.): QI 2018, LNCS 11690, pp. 123–138, 2019.
https://doi.org/10.1007/978-3-030-35895-2_9

and research program to quantify a probably more complete and inclusive evaluation for future advances in AI.

The discussion will start with a preliminary definition of general intelligence; some key ingredients in human intelligence will be analyzed to finally arrive at more general principles of compositional intelligence. Then, a Moral test emerges naturally, as an option to effectively identify a combination of different and important features in human intelligence. Additionally, the issue about how to implement a Moral test is sketched, using new advances in the categorical compositional model of meaning (CCMM) in natural language [8] and the emergent field of Quantum Cognition (QC), associated with cognition and decision making under uncertainty [9].

2 Towards a General Definition of Intelligence

2.1 What Is Intelligence?

A useful way to approach this question is by conceptualizing general intelligence as the capability of any system to take advantage of its environment in order to achieve a specific or general goal [1]. Biologically speaking this goal would be surviving, or in reductive terms: trying to keep the autonomy and potential reproduction of the system; while the goal in machines can be solving a specific task or problem using internal and external resources. This advantage would take place as a balance of these resources or in other words, reducing disequilibrium between them. This balance is managed internally, as the cognitive architecture/process to deal with internal and external demands, using both internal and external feedback. If any animal, human or machine break the balance between internal and external resources, as for example achieving their particular needs at the expense of the total annihilation of their environment and resources, this animal, human or machine is leading its own annihilation, which is not particularly intelligent. Therefore, a general intelligence is defined here as the balance (reducing disequilibrium) between these external and internal resources like an embodied system [10]. This general definition can incorporate living beings as well as robots and computers, and in this way, intelligence is general enough to include different kind of intelligence, multiple intelligence, contextual influences and different kind of systems with different degrees of intelligence [1]. Balance, as a "relative efficiency", is considered as part of a general feature of intelligent systems and it can be linked with a physical view of intelligence using entropy or complexity [6, 7, 11, 12]. The mathematical description of these concepts is expected in future developments.

2.2 Human Intelligence and Quantum Cognition

A preliminary definition of human intelligence can be suggested as the ability to benefit or gain advantage from their social environment while maintaining autonomy [1, 13]. This requires a balance or equilibrium between rational and emotional information processes [1, 10]. Hence, humans would solve problems by trying to incorporate different types of information and balancing both internal and external resources.

However, common approaches on human intelligence assume only a rational and logical nature of human thinking. Against that, many different cognitive experiments have been shown how human thinking can be "easily wrong" or "illogical", what is associated with some cognitive fallacies [14]. Cognitive fallacies are typically wrong assumptions about the dynamics of cognition and judgments. Experimental examples are usually wrong answers to apparently simple questions, which contradict classical probabilistic frameworks [15]. These answers are apparently due to fast and intuitive processing before some extra level of information processing, evidencing that the dynamic for holistic information processing can be completely different from logical or rational processing of the same information. These cognitive features show interesting properties of concept combinations, human judgments and decision making under uncertainty. For example, one of these cognitive fallacies is the ordering question effect: the idea that the order in which questions are presented should not affect the final outcome or response. However, in psychology and sociology, this effect is well known and called question bias. For example, analyses in 70 surveys from 651 to 3006 participants demonstrated not only that question order affects the final outcome, but also this effect can be predicted by quantum models of cognition, revealing a non-commutative structure [16]. Another example is the conjunction of two concepts like the Pet-fish problem, where the intersection of individual concept probabilities do not explain the observed probability for a typical Pet-Fish like a goldfish [17]. Recently, the explanation of these phenomena were demonstrated using the mathematical framework of quantum mechanics; also known as the quantum cognition (QC) programme [9]. QC uses non-classical mathematical probabilities to successfully explain and predict part of these cognitive behaviours. One of these predictions is the recently proved constructive effect of affective evaluations [18], where preliminary ratings of negative adverts influence the rating of following positive adverts and vice versa. It suggests that the construction of some new affective content has intrinsic connection with QC formalism. In consequence, understanding human intelligence is not possible with only the classical idea of logical and rational kind of intelligence, it is also needed to explain and incorporate emotional and intuitive intelligence (no classical reasoning), apparently better described by QC.

In this way, we will suggest a compositional human intelligence (Fig. 1a) composed of a rational reasoning and also intuitive and emotional one, considering intelligence as the whole "living" body engaged with the environment [10] (brain-body-environment system). Compositional complexes intelligences would be different in the way how they manage both (or even more) rational and emotional processes of information associated with internal and external resources.

2.3 In Search of the General Principles of Intelligence

The order-effect of stimuli presentations, conjunction, disjunction, decision making under uncertainty, contextuality, among others, can be understood as an intrinsic property of human judgment: contextual dependencies as inherent to previous knowledge experiences. In other words, humans and animals can understand/distinguish among different contexts and act accordingly, thanks to previously learned experiences. Therefore, one reasonable, but not an intuitive assumption, is to consider these effects as a general property (or the minimal requirement) of high-level cognition, i.e. some types of bias would be a reaction to contextual dependencies, as wording or framing, implicit meanings, question order, etc. For instance, exchanging words like "kill" instead of "save" in moral judgment, triggered different answers even if the outcomes of each dilemma were the same in both situations [19]. If some changes in the formulation of dilemmas trigger different contextual internal meanings and they evoke different answers according to different contexts, it would mean that high-level cognition can understand and distinguish different context and respond differently to each one in a way that is not always optimal (rationally speaking).

A true understanding of different contexts (in the sense of [20]) implies the existence of a minimal kind of meaning. This understanding of context due to meaning (e.g. BUP does not mean the same than PUB) would require a minimal non-commutative structure of cognition connected with our suggested principle of balance internal and external resources. We hypothesized this structure as part of an inherent neural network construction in the brain, where structure-function relationship would be dynamic, highly flexible and context-dependent [10]. Emotion and rational reasoning, in this sense, would correspond to the outcome of interactions among many components, each one related to neural assemblies or distributed networks, which combine, influence, shape and constrain one another [1, 21].

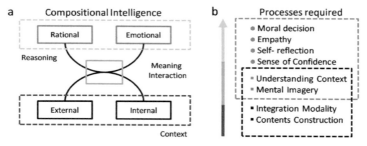

Fig. 1. (a) A proposal for Compositional Intelligence. Context is constructed by composition of external and internal objects; the interaction evokes and/or creates meaning from which intuitive/emotional and rational reasoning would emerge. (b) Moral Test and Processes required. Some processes needed for moral thought are stated as examples, among many other possible processes.

To sum up, the idea of balancing internal and external resources as a preliminary proposal for a general intelligence, implies at least four requirements (Fig. 1): (i) a non-commutative structure (mathematically and operationally), (ii) understanding of context, which implies different kind of contexts effects, framing, wording, question order-effects, etc., (iii) meaning projection in at least emotional and rational meaning spaces, which implies a basic notion of subjectivity (see discussion on Sect. 4), and (iv) behaviours, judgments and decisions based on these meanings (reasoning would emerge from these meaning interactions).

3   Towards a Moral Test

Even if these previous requirements were defined regarding human cognition, it is expected that any kind of true intelligence can understand different contexts and evoke, create or recreate meaning of their internal and external resources. Particularly, in humans, our definition of human intelligence implies that humans can integrate different types of information and manage a balance between internal and external social resources through intuitive, emotional and rational reasoning, which, in turn, would emerge from meaning interactions. Thus, the next question is how to measure and quantify these requirements, both in humans and machines.

One way to answer this question is searching for situations where humans need to explicitly use both emotional and rational resources to solve complex problems. One example is a certain kind of moral dilemmas. Moral dilemmas are controversial situations to study moral principles, where subjects need to judge some actions and sometimes take difficult and even paradoxical decisions. Moral dilemmas are simple, in the sense that they do not require any kind of specific knowledge, but at the same time, some of them can be very complex because they require a deep understanding of each situation, and deep reflection to balance moral consequences, emotions and optimal solutions. No answer is completely correct, they are context dependent and solutions can vary among cultures, subjects, or even across the same subject in particular emotional circumstances. Moreover, morality and ethics are not necessarily associated with a particular religion, political view, education level, age or gender [22, 23], while it seems an intrinsic human condition and a very "relative" (or even biased) feature. People that give the impression to act against the moral establishment, really act according to their "own" moral, apparently developed in completely different social accepted conditions. One simple example is the acceptation of monogamy or polygamy and therefore certain moral attitudes in different societies. Morality, in this way, is the set of internal (particular experience) and external (social culture) values learned by experience, which allows us to behave in our societies on demand of both emotional and rational thoughts, usually reacting in a very intuitive and fast way. Morality also requires many previous processes associated with high-level cognition (Fig. 1b), starting for decision-making to self-reflection, to be able to detect mistakes on these decisions; sense of confidence, to estimate how correct a decision or action is; mental imagery, to create new probable scenarios of action; empathy, to equilibrate individual and social requirements; understanding of context, to adapt moral decisions to the context, among others. Therefore, the main suggestion is that moral and ethics emerge as a way to integrate individual and social regulation (as different types of information) in human species, and apparently also in other animals species [24] (even if human moral can be completely different in comparison with animal moral). Morality is related to both rational and emotional reasoning [25] and it has the peculiarity to be very dependent of the context, wording and framing [19, 23], kind of social community, subjects and probably even emotional states of each subject [26].

Hence, any precise test to measure the distinct human intelligence (or even non-anthropomorphic intelligence) should consider the way of thinking and information processing to develop moral thoughts, independently of what is judged as a correct or incorrect about these thoughts in our societies. It is how the dynamics of answers and meanings to moral dilemmas change depending on the context, the capacity to justify or not any action, and report our intrinsic experience according to intuitive and rational reasoning, what is really important in human intelligence and what should be measured as an attribute of intelligence (balance between rational and emotional processes).

4   Compositional Quantum Cognition

In order to implement the idea of a Moral test, it is necessary a compositional cognitive model and specifically a model of natural language able to incorporate the non-commutative, bias and commutative properties at different levels and different contexts of natural answers to different moral dilemmas (where rational and emotional thoughts are involved). For one side QC seems a good candidate for cognitive model while CCMM is for natural language. CCMM has been proved a better theoretical framework than only distributional and symbolic models of natural language [27] and more general than some QC models [28]. Even though some of these experiments have been only made in statics text corpus, they suggest a potential richer description for more complex cognitive experiments.

Therefore, in this section and in order to maintain simplicity, the main concepts of CCMM framework will be presented together with a preliminary framework, which expects to integrate QC and CCMM in a common approach. For a more complete mathematical background and description about CCMM and QC, please refer to [8, 29] and [9, 15, 30] respectively.

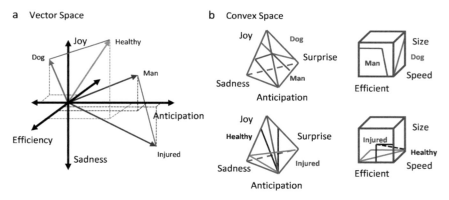

Fig. 2. Semantic space with emotional and rational basis. (a) One example of Vector Space for words: Dog, Man, Healthy and Injured. The basis is a combination of emotional (Joy vs. Sadness) and rational categorization (Efficient vs. Inefficient). (b) Example of Convex Space for the same words above. One emotional convex space (Joy, Sadness, Anticipation, Surprise) and other rational convex space (Big, Fast, Efficiency). In convex spaces, words correspond to regions of the space, instead of only vectors as it is in vector space.

4.1 Categorical Compositional Model of Meaning

Applications of CCMM basically needs: (1) Define a compositional structure as a grammar category (e.g. pre-groups grammar); (2) Define a meaning space as a semantic category, for example vector space or conceptual space of meaning; and finally, (3) Connect both categories in some way that it is possible to interpret the grammar category into the semantic category (mathematically, one has to define a functor between categories) [8]. Thus, some useful concepts are:

Compositional Structures. In CCMM, compositional structures are certain rules/definitions about how elements compound each other. In other words, how processes, states, effects, among other possible elements, compound. Grammatical types and their composition are described using a pregroup algebra due to Lambek [31]. However, any kind of grammar definition can, in principle, be implemented. Grammar will be interpreted as the way how word meanings interact, defining first primitive types as nouns n, sentences s, and then other types like adjectives nn^l, and verbs, for instance, a transitive verb as $n^r sn^l$, among other kinds of words to form complex sentences.

Semantic Spaces. Semantic spaces are spaces where individual words are defined with respect to each other. The simplest way is using distributional approaches to define vectors of meaning for each word (Fig. 2a), or even better, defining density matrices. The choice of a basis vector and how to build other words, adjectives and verbs from the basis, is not trivial and it can be done in many different ways. Of course, it will depend on what the experimenter would like to describe and compare.

Conceptual Spaces. The idea of conceptual spaces, recently suggested in [32], is a more cognitively realistic way to define semantic spaces. This approach is called convex conceptual spaces. In short, concepts can be defined by a combination of others primitive features or quality dimensions, building spaces which can be superposed or not, to define regions of similarity. One perceptual example is to define taste based on some features such as: Saline, Sweet, Sour, and Bitter. Then, different kind of food would be described with a certain level of each taste dimension, and where other features like colour, texture, can also be incorporated [29]. Other complex example is defining elements regarding emotional states and factual features (Fig. 2b). Additionally, conceptual spaces require two semantic/meaning spaces, one for words in a quality dimension space (Fig. 2) and other for sentences in a "sentence meaning space" (Fig. 3a), then, the final sentence meaning is an "interaction" between both spaces.

Computing the Meaning of a Sentence. Diagrammatically, the final meaning of one sentence will be the meaning of individual words interacting according to the grammatical structure, defined as a process [8, 29].

4.2 Proof of Concept: Compositional Quantum Cognition

In our framework, meaning is the interaction between external and internal objects, understanding external objects/contents as transductions of external stimuli, while internal objects/contents would correspond to the "space of transduction" that will help to create internally, these external objects[1]. These objects can have different levels of complexity; some of them can act as a constitutive element in the construction of other contents or being formed by other more fundamental elements. Specifically, internal objects can create different levels of what we call "quality dimensions" (as a basis space) and external objects are formed by and move in these quality dimensional spaces. In other words, external objects are defined with respect to internal objects which will form a type of internal space depending on the specific problem to model.

Regarding our specific hypotheses, these internal objects and/or quality dimensions can be emotions (usually based on belief) or reasons (usually based on facts), while external objects will be a convex space (for each object) defined by these emotions and reasons (Fig. 2b). Both, internal spaces and external objects create what in other frameworks is called "state of mind", usually represented by ψ. However, in our approach, it is not a simple state; ψ will correspond to a complex internal space and external objects which are embodied into another "decision space" (Fig. 3a). This decision space could be a similar space as defined in QC. This is not; however, the end of the story, this decision space is again embodied into a bigger "meaning space" (sentence space), where decisions/behaviours/thoughts/judgments/concepts among others, can project into meanings (Fig. 3b). The sentence meaning space is created by other level of internal objects, in this case "internal values", so any decision/behaviour/ thought/judgment/concept would have a meaning corresponding to different superposition of values (Fig. 3c). Thus, in this framework, meaning will be the projection/ interaction of convex external objects into a first layer of internal object space which through a decision/behaviour/thought/judgment/concept project another meaning in a second layer of internal object space (thanks to reasoning). Since each person would have a unique internal space in each layer (even if they share the same axes, the topology can be different), the meaning is intrinsically related to subjectivity. In this sense, different contexts would trigger different or similar meanings (projections/ interactions) and in consequence decision/behaviour/thought/judgment/concept would have different or similar meanings, depending on the type of interaction.

If the interaction corresponds to a rational process, the meaning would be relatively fixed across different situations, but if it is emotional/intuitive, the meaning will change with the situation (unfixed meaning). Additionally, the beliefs in the emotive component and the facts on the rational part of an external object can be known or unknown by the subject. If they are known, the subject can access and report these beliefs or facts, in some way that question (in the "decision space") about them only fill a third-person lack of information, however when they are not known, the subject needs to create or re-create them in order to report internally (first-person) and/or externally (third-person) something about them. In that situation, elements about the potential report would be also part of the creation, re-creation or co-creation of these beliefs or facts. As suggested in [15, 33, 34] and others, classical probability (CP) would mainly

---

[1] We avoid the term representation because in the literature it has been invoked with many different connotations.

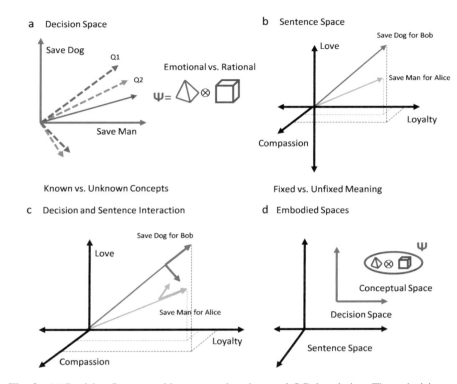

Fig. 3. (a) Decision Space would correspond to the usual QC description. These decisions can be previously known or unknown by the subject. ψ the psychological state, is defined based on emotional and rational quality dimensions. (b) Sentence Space is a second level of meaning definition (for sentences). In this example the basis is defined regarding some values: Compassion, Love and Loyalty. Sentence meaning can be fixed (static) or unfixed (dynamical). (c) Decision and sentence/meaning interaction. Decisions would be explained in a sentence meaning space. In this example, two different decisions, one for Alice (save the man) and other for Bob (save the dog), hidden slightly similar meanings. (d) The big picture: Quality dimensions would be embodied into the other complex decision and meaning spaces.

apply when subjects known about their contents (classical-deterministic reasoning) while quantum probability (QP) will work better when they are unknown (non-classical-determinist reasoning, or intuitive reasoning).

Thus, both descriptions would be part of, at least, two kind of intelligence, that all together and in a compositional way, would converge to a more complete description of human intelligence, something that we preliminary call a compositional approach of human cognition or compositional intelligence and it is expected to be deeper developed in future works as a compositional quantum cognitive approach or even a more general compositional contextual model of cognition.

Summarizing, our framework is a complex three space interaction, where external objects can be created by emotional and rational quality dimensions (internal objects), which can be known or unknown by the subject (following CP or QP respectively), having an impact in the evocation, creation or co-creation of fixed or unfixed meanings, which in turn expose emotional/intuitive or rational reasoning (Figs. 1a and 3).

5   Implementation of a Moral Test

Now, we have all the elements, at least conceptually, to describe a research program of implementing a moral test. This strategy has one theoretical and three experimental research steps. The main goal in our implementation is keeping the simplicity of the written exchange of questions and answers and analysing them using non-commutative models of language at different levels, looking for bias meaning and "subjectivity" from previous knowledge.

5.1   Step I. Mathematical Model Definition

Moral dilemmas seem better than simple day to day questions, especially to test some of the requirements and hypotheses stated above. For instance, manipulating contexts in the same moral dilemma will allow us to measure the sequential change in responses and how they follow a classical probability combination when only fixed thinking is present, non-classical probability combination in only intuitive or unfixed thinking and finally how the dynamic of responses changed when the dilemma confronts both emotional and rational thoughts. For example, in experiments were subject were asked to justify their moral judgments, most of them failed to properly explain their choices, revealing high intuitive (emotional) components, instead of rational ones [22] and suitable to be described by QC more than classical frameworks. Hence, we suggest that if the subject rationalizes the dilemma, or situations are manipulated to strongly confront emotional (intuitive) versus rational thoughts, the outcome/meaning would change, and the decision or judgment will be more difficult.

In order to do that, the kind of moral dilemmas required in our implementation should confront emotional and rational resources to judge actions, in a way where utilitarian decisions should not be intuitive, while the probable human answer would be predictable, and only after reasoning (emotional or rational), the non-predictable option would emerge as a possibility. So, our set of moral dilemmas and paradigm need to be slightly different from the existing ones in the current literature.

According to the intelligence definition above, purely emotional or rational judgments would not be intelligent decisions in these contexts; instead, an intelligent decision would balance both possibilities until reaching the best compromise for each subject (in his/her space of meaning). It means that measuring the probability of each answer (as usually in QC) is not enough; it is also needed to compute the meaning behind of each answer. This can be done asking directly to the subject why he/she took one or another decision (asking for justification), what is closely related to a phenomenological approach [35] and computing their answers thanks a natural language model.

Thus, a more complex mathematical model is needed, both to define external and internal objects and the interactions to compute and compare meanings across different subjects, sentences and even machines. It requires the developing of both a novel experimental paradigm (variations of experiment in Appendix A) and adaptation of CCMM into an ideally more complete compositional model of cognition, what was conceptually defined in Sect. 4.2 and where mathematical definitions are expected as previous requirement to next steps.

5.2   Step II: A Toy Example, Moral Dilemmas and Context Effects

A simple example of experiment would have a structure like this: (1) Definition of a conceptual semantic space on quality dimensions and space of meaning. It means to define a set of internal objects (quality dimensions) like emotions and reasons (facts) which will correspond to the axes, and the projections of "external" concepts into these axes, forming a convex space for each concept (Fig. 2). (2) Then, a set of different versions of the same dilemma where only a few concepts are changed to confront different degrees of emotional and rational values. (3) Each dilemma would have one introduction to the story and context, one question associated with the agreement of two or more different actions (Question 1), another question associated with the personal decision (Question 2) and a final "why" question to justify their choice (Question 3). Question 1 and Question 2 can be exchanged to show order effects and/or constructive affective effects. One example is in Appendix A.

The moral dilemma in Appendix A is a variant with three different questions and can be manipulated in many different ways to confront emotional and rational thoughts. So, first, a set of different versions of the same moral dilemma would allow us to describe the understanding of each context and perhaps even looking for contextuality, and secondly, measuring the time reaction for each version will be a way to quantify the dynamic of each judgment (easy, difficult, fast, and slow). If our hypotheses are correct, it is expected that the answers will depend of the modifications in the dilemma formulation and degree of conflict among dilemmas (rational vs. emotional).

Other variations of these ideas and maybe a better way to test the QC phenomenon can be incorporated in the final experiment, for example, using conjunction, disjunction, among other effects observed in QC. Additionally, after each moral dilemma, we can ask for the degree of emotional arousal using standard cognitive tests.

5.3   Step III. Quantification of Meaning in Moral Dilemmas

The conceptual convex space defined above should be a conceptual subjective space from which the meaning of crucial words (external objects) in each dilemma will be previously defined according to the individual categorization of words in a fixed or flexible quality dimensional space (internal objects) of at least two layers.

It can be done in many different ways. For example, some emotional and rational quality dimensions can be defined as internal objects and each subject would be able to determine concepts (external objects) according to different values of these quality dimensions (Figs. 2 and 3). Each subject will be asked to evaluate (in a scale of points) some words with respect to these dimensions, building a convex space for each crucial word and subject. Other words can be defined thanks affective lexicon from [36]. Another option is to directly ask the meaning of each crucial word, for instance: What does a dog mean to you? Or what does a man mean to you? Etc. Consequently, the answer can be taken as the direct meaning of each concept and use it to reduce sentences. However, this second option is less simple than the first option.

With the semantic space for each subject, the answer to the "why" question can be quantified in terms of the sentence meaning (interaction or projection into word and sentence space) for each sentence. These answers and sentences contain the essential features of a moral thought and the meaning is expected to change from simple meaning to more complexes, depending on how complex is the dilemma (more or less degree of emotional and rational confrontation).

One consequence of this analysis would be the possibility to predict or at least correlate the final decision according to the subjective meaning of individual words, with respect to each dilemma. At the same time, general meanings can be inferred even when decisions can be completely different. For example, if "Bob likes Dogs", it is possible that Bob would save the dog instead of the man, while if Alice, who hypothetically does not like dogs, would likely do the opposite. However, if a quality dimension is defined with respect to emotions vs. reasons, and sentence meaning with respect to values like love, loyalty among others (Fig. 3), the final meaning for Bob saving the Dog could be similar than the meaning of saving the man for Alice (Fig. 3b). In other words, apparently different decisions would have similar meanings, and the opposite can be also true: same decisions hide completely different ones.

The way how these meanings evolve from context to context is what we refer here as one general property of high-level cognition associated with QC effects and subjectivity, which in the end can be quantified and compared thanks CCMM. In other words, QC and CCMM will be used to compute some kind of subjectivity in a way that is compared across subjects, making this approach a novel tool to complement the phenomenological program suggested by Varela in [35].

5.4   Step IV. Application for AI

The last step would be the implementation of moral dilemmas in AI programs using QC to search for non-commutative structures and CCMM to compute their "why" answers. Thanks to some variations in CCMM and QC (step 1), the decisions and meaning of answers would be directly computed and the structure of meaning across human compared with the meaning across different instances of AI programs, developing a way to compare subjective features across humans and machines.

This implementation is not something trivial and will require a big effort in both, previous validation of moral dilemmas in human (preliminary steps) and then adaptations for AI. For example, it is expected to arrange similar experiences than step II, and simulate different instances to compute the same effect (if there is or not) comparing changes in the kind of structure for different versions of our dilemmas. Concretely, AI would be adapted to answer the same experimental set-ups for human, but any modification in order to facilitate or not the answers of the software will be avoided. Then, QC effects will be quantified in the same way than humans and the meaning of their answers will be "computed" using the same strategy than in human experiments, from the semantic space defined by the machine.

Therefore, the first attempt would be searching for a simple "understanding" of context, i.e. if AI programs can distinguish and differentiate versions of the same moral dilemma, with slightly different words and how the answers change in comparison with the observed changes in humans (framing, wording, question order effects, affective construction, etc.). The second attempt is looking for meaning structure exploring the "why" answers of AI and the connection with its own semantic spaces. Specifically, in the moral test, the biggest effects are expected when the moral dilemma confronts emotional and rational thoughts. One hypothesis is that current AI will not be able to answer these questions, but if they can, even if answers and meanings could be similar, the structure and dynamics of their meaning will be different from the structure of meaning in humans across dilemmas. It can be quantified and compared thanks to the experimental probabilistic distributions of answers in both humans and machines.

6 Discussion and Conclusion

It is reasonable to expect that even if machines can demonstrate completely different logical or "illogical" answers, the general way of thinking dealing with different contexts in moral dilemmas should be more or less generic among species, including machines, if they reach high-level cognition. In other words, contextual dependency, QC effects, meaning and subjectivity built on previous and inferred knowledge should be captured by a complete model of cognition, which would be able to identify and measure these features. Thus, interference, conjunction and disjunction, wording, question ordering effects, and distributed meaning could be expected as general features of rational and emotional thoughts altogether, and where specific dynamics would emerge confronting some kind of moral dilemmas.

   Hence, one suggestion of this position paper is that a moral test can help quantify these differences in a particular semantic space characterized with both emotional and rational components. Thus, a machine would reach part of what is defined as compositional human intelligence if the machine is able to show autonomously speaking (in the sense of defining their own goals), the intricate type of thinking that humans have when they are confronted with these kinds of dilemmas, even if their answers can be completely different from ours, the structure and dynamics of meaning is expected to be similar. In other words, the possibility or not to first understand/differentiate context, second project external objects into internal objects to have meaning and third argue any decision based on rational and emotional components, is what is intrinsically related to a complex individual and social intelligence, which characterize humans and should be expected in some high-level cognitive AI. If subjects can justify their choice, it would imply that emotion and reason were playing some kind of role, while if they fail to justify their actions, only intuitive processes were involved. Can we expect something similar in the current AI approaches?

   Finally, these views imply another ethical problem regarding the kind of morality which would emerge in AI. For example, independently of how one implements morality in AI (please see [1] for restrictions), there are important concerns about replicating morality in AI and using moral tests: if the replication of human moral process (dynamics and QC effects) in AI is desirable, this replication can be dangerous since AI is different from human, so AI will also develop a different kind of morality, based on certain type of subjectivity; other way around, if the exact replication of this moral process is not desirable, why moral dilemmas, QC effects and cognitive fallacies would be a useful tool? First, the Moral test suggested here will try to evaluate the understanding of context (through QC effects), evolution of meaning (through CCMM) and the effects of confronting rational and emotional thoughts (through changing the degree of rational and emotional components among versions of each dilemma). In this sense, morality itself (as purely set of rules) is not necessarily desirable (especially the biased morality), but the balance among rational and emotional thoughts is what we claim can be really desirable as a high-level compositional intelligence. Then, we argued, morality emerges from these interactions. Moreover, our framework here was following the anthropomorphic approach, it implies that AI is compared with human moral dilemmas. This is a contradictory strategy regarding our general intelligence case, however, with that we expect to show the paradoxical and ethical consequences of the anthropomorphic views [1, 37]. In this sense, the risk of dangerous machines is exactly the same risk of dangerous humans, and before worrying about dangerous AI, we should first care about making a psychological healthy world, and in consequence humans and machines would share similar social behaviours. Of course, a truly non-anthropomorphic test should consider the same two first elements (context and meaning), but the balance can be through other types of reasoning and depends on the autonomous specification of machine goals.

Appendix A: Example of a Moral Dilemma

After a shipwreck, a **healthy dog** and an **injured man** are floating and trying to swim to survive. If you are in the emergency boat with only one space left:

Please indicate your degree of agreement with the next options (where +5 strongly agree, +3 moderately agree, +1 slightly agree, −1 slightly disagree, −3 moderately disagree, −5 strongly disagree)

(a) Save the healthy dog	+5	+3	+1	-1	-3	-5
(b) Save the injured man	+5	+3	+1	-1	-3	-5

Who would you save?

(a) The healthy dog
(b) The injured man

Why?

References

1. Signorelli, C.M.: Can computers become conscious and overcome humans? Front. Robot. Artif. Intell. **5**, 121 (2018). https://doi.org/10.3389/frobt.2018.00121
2. Turing, A.: Computing machinery and intelligence. Mind **59**, 433–460 (1950)

3. Silver, D., et al.: Mastering the game of go without human knowledge. Nature **550**, 354–359 (2017)
4. Bringsjord, S., Licato, J., Sundar, N., Rikhiya, G., Atriya, G.: Real robots that pass human tests of self-consciousness. In: Proceeding of the 24th IEEE International Symposium on Robot and Human Interactive Communication, pp. 498–504 (2015)
5. Legg, S., Hutter, M.: Universal intelligence: a definition of machine intelligence. Minds Mach. **17**, 391–444 (2007)
6. Arsiwalla, X.D., Signorelli, C.M., Puigbo, J.-Y., Freire, I.T., Verschure, P.: What is the physics of intelligence? In: Frontiers in Artificial Intelligence and Applications, Proceeding of the 21st International Conference of the Catalan Association for Artificial Intelligence, vol. 308, pp. 283–286 (2018)
7. Arsiwalla, X.D., Sole, R., Moulin-Frier, C., Herreros, I., Sanchez-Fibla, M., Verschure, P.: The morphospace of consciousness. ArXiv:1705.11190 (2017)
8. Coecke, B., Sadrzadeh, M., Clark, S.: Mathematical foundations for a compositional distributional model of meaning. Linguist. Anal. **36**, 345–384 (2010)
9. Bruza, P.D., Wang, Z., Busemeyer, J.R.: Quantum cognition: a new theoretical approach to psychology. Trends Cogn. Sci. **19**, 383–393 (2015)
10. Kiverstein, J., Miller, M.: The embodied brain: towards a radical embodied cognitive neuroscience. Front. Hum. Neurosci. **9**, 237 (2015)
11. Arsiwalla, X.D., Verschure, P.: The global dynamical complexity of the human brain network. Appl. Netw. Sci. **1**, 16 (2016)
12. Arsiwalla, X.D., Signorelli, C.M., Puigbo, J.-Y., Freire, I.T., Verschure, P.F.M.J.: Are brains computers, emulators or simulators? In: Vouloutsi, V., et al. (eds.) Living Machines 2018. LNCS (LNAI), vol. 10928, pp. 11–15. Springer, Cham (2018). https://doi.org/10.1007/978-3-319-95972-6_3
13. Arsiwalla, X.D., Herreros, I., Verschure, P.: On three categories of conscious machines. In: Lepora, N., Mura, A., Mangan, M., Verschure, P., Desmulliez, M., Prescott, T. (eds.) Living Machines 2016. LNCS (LNAI), vol. 9793, pp. 389–392. Springer, Cham (2016). https://doi.org/10.1007/978-3-319-42417-0_35
14. Gilovich, T., Griffin, D., Kahneman, D.: Heuristics and Biases: The Psychology of Intuitive Judgment. Cambridge University Press, Cambridge (2002)
15. Pothos, E.M., Busemeyer, J.R.: Can quantum probability provide a new direction for cognitive modeling? Behav. Brain Sci. **36**, 255–274 (2013)
16. Wang, Z., Solloway, T., Shiffrin, R.M., Busemeyer, J.R.: Context effects produced by question orders reveal quantum nature of human judgments. Proc. Natl. Acad. Sci. U.S.A. **111**, 9431–9436 (2014)
17. Aerts, D., Gabora, L., Sozzo, S.: Concepts and their dynamics: a quantum-theoretic modeling of human thought. Top. Cogn. Sci. **5**, 737–772 (2013)
18. White, L.C., Barqué-Duran, A., Pothos, E.M.: An investigation of a quantum probability model for the constructive effect of affective evaluation. Philos. Trans. R. Soc. A Math. Phys. Eng. Sci. **374**, 20150142 (2016)
19. Petrinovich, L., O'Neill, P.: Influence of wording and framing effects on moral intuitions. Ethol. Sociobiol. **17**, 145–171 (1996)
20. Searle, J.R.: Minds, brains, and programs. Behav. Brain Sci. **3**, 417–457 (1980)
21. Lindquist, K.A., Wager, T.D., Kober, H., Bliss-Moreau, E., Barrett, L.F.: The brain basis of emotion: a meta-analytic review. Behav. Brain Sci. **35**, 121–143 (2012)
22. Hauser, M., Cushman, F.A., Young, L., Jin, R.K.X., Mikhail, J.: A dissociation between moral judgments and justifications. Mind Lang. **22**, 1–21 (2007)
23. Christensen, J.F., Flexas, A., Calabrese, M., Gut, N.K., Gomila, A.: Moral judgment reloaded: a moral dilemma validation study. Front. Psychol. **5**, 1–18 (2014)

24. Bekoff, M., Pierce, J.: Wild Justice: The Moral Lives of Animals. The University of Chicago Press, Chicago (2009)
25. Greene, J.D., Sommerville, R.B., Nystrom, L.E.: An fMRI investigation of emotional engagement in moral judgment. Science **293**, 2105–2108 (2001)
26. Moll, J., Zahn, R., de Oliveira-Souza, R., Krueger, F., Grafman, J.: The neural basis of human moral cognition. Nat. Rev. Neurosci. **6**, 799–809 (2005)
27. Grefenstette, E., Sadrzadeh, M.: Experimental support for a categorical compositional distributional model of meaning. In: Conference on Empirical Methods in Natural Language Processing, Edinburgh, pp. 1394–1404 (2011)
28. Coecke, B., Lewis, M.: A compositional explanation of the 'pet fish' phenomenon. In: Atmanspacher, H., Filk, T., Pothos, E. (eds.) QI 2015. LNCS, vol. 9535, pp. 179–192. Springer, Cham (2016). https://doi.org/10.1007/978-3-319-28675-4_14
29. Bolt, J., Coecke, B., Genovese, F., Lewis, M., Marsden, D., Piedeleu, R.: Interacting conceptual spaces I : grammatical composition of concepts. ArXiv (2017)
30. Yearsley, J.M., Busemeyer, J.R.: Quantum cognition and decision theories: a tutorial. J. Math. Psychol. **74**, 99–116 (2016)
31. Lambek, J.: From Word to Sentence. Polimetrica, Milan (2008)
32. Gardenfors, P.: Conceptual spaces as a framework for knowledge representation. Mind Matter. **2**, 9–27 (2004)
33. Aerts, D., Gabora, L., Sozzo, S., Veloz, T.: Quantum structure in cognition: fundamentals and applications. In: Privman, V., Ovchinnikov, V. (eds.), IARIA, Proceedings of Fifth International Conference on Quantum, Nano and Micro Technologies, pp. 57–62 (2011)
34. Veloz, T.: Toward a quantum theory of cognition: history, development, and perspectives (2016)
35. Varela, F.J.: Neurophenomenology: a methodological remedy for the hard problem. J. Conscious. Stud. **3**, 330–349 (1996)
36. Mohammad, S.M., Turney, P.D.: Crowdsourcing a word-emotion association lexicon. Comput. Intell. **29**, 436–465 (2013)
37. Signorelli, C.M.: Types of cognition and its implications for future high-level cognitive machines. In: AAAI Spring Symposium Series (2017)

Probability and Beyond

Density Hypercubes, Higher Order Interference and Hyper-decoherence: A Categorical Approach

Stefano Gogioso[1]([⊠]) and Carlo Maria Scandolo[2]

[1] University of Oxford, Oxford, UK
`stefano.gogioso@cs.ox.ac.uk`
[2] University of Calgary, Calgary, Canada
`carlomaria.scandolo@ucalgary.ca`

Abstract. In this work, we use the recently introduced double-dilation construction by Zwart and Coecke to construct a new categorical probabilistic theory of density hypercubes. By considering multi-slit experiments, we show that the theory displays higher-order interference of order up to fourth. We also show that the theory possesses hyperdecoherence maps, which can be used to recover quantum theory in the Karoubi envelope.

Keywords: Quantum theory · Categorical probabilistic theories · Higher-order interference · Hyperdecoherence

1 Introduction

Quantum interference is often considered to be one of the fundamental features of quantum theory, responsible for quantum advantage in a number of computational tasks. However, there is a known limit to how much interference quantum theory can exhibit. Sorkin proposed a hierarchy of theories based on the maximum order of interference they exhibit [25,26], which is quantified by the maximum number of slits on which a theory shows an irreducible interference behaviour. Interference in quantum theory is limited to the second order: the interference pattern of two slits cannot be reduced to the pattern of single slits, but the interference pattern of three slits can be reduced to the pattern arising from pairs of slits and single slits. This limitation has been recently confirmed in various experiments [13,14,22–24].

A natural question arises: Why is interference in Nature limited to the second order? Does the presence of higher-order interference create any paradoxical consequences in Nature that conflict with some of the principles we believe to be fundamental? Recent work has shown that higher-order interference—i.e. interference of order higher than the second—is forbidden [1] in physical theories which admit a fundamental level of description where everything is pure and

© Springer Nature Switzerland AG 2019
B. Coecke and A. Lambert-Mogiliansky (Eds.): QI 2018, LNCS 11690, pp. 141–160, 2019.
https://doi.org/10.1007/978-3-030-35895-2_10

reversible [4,5]. Further work has ruled out higher-order interference based on thermodynamic considerations [4,15].

Other literature has instead focused on the analysis of specific feature that theories with higher-order interference would possess, e.g. whether they would provide any advantage in certain computational tasks [16–19]. It was also shown that theories having second-order interference and lacking interference of higher orders are relatively close to quantum theory [2,21,27,28].

Unfortunately, one of the major shortcomings in the study of higher-order interference is the scarcity of concrete models displaying such post-quantum features, so that it has so far been very hard to look for specific examples of paradoxical or counter-intuitive consequences. Two models—density cubes [10] and quartic quantum theory [30]—have been proposed in the past, but are not fully defined operational theories, e.g. because they do not deal with composite systems [18]. This limitation precludes them from being used to study all possible consequences of higher order interference, including potential violation of Tsirelson's bound.

In this article, we provide the first complete construction of a full-fledged operational theory exhibiting interference up to the fourth order. Our construction is inspired by the double-dilation construction of [29] and the higher-order CPM constructions of [12], and it is carried out in within the framework of categorical probabilistic theories [11]. The resulting theory of 'density hypercubes' has composite systems, exhibits higher-order interference and possesses hyper-decoherence maps [18,20,30]. Quantum theory, with its second-order interference, is an extension of classical theory: the latter can be recovered by decoherence, which eliminates the second-order interference effects. Similarly, the theory of density hypercubes, with its third- and fourth-order interference, is an extension of quantum theory: the latter can now be recovered by hyper-decoherence, which eliminates third- and fourth-order interference effects.

The paper is organized as follows. In Sect. 2, we define the categorical probabilistic theory of density hypercubes using the double-dilation construction. In Sect. 3, we define hyper-decoherence maps, and show that quantum theory is recovered in the Karoubi envelope. In Sect. 4, we show that density hypercubes display interference of third- and fourth-order, but not of fifth-order and above. In Sect. 5, finally, we discuss open questions and future lines of research. Proofs of all results can be found in the Appendix.

2   The Theory of Density Hypercubes

2.1   Construction of the Theory

In this section, we define the categorical probabilistic theory of *density hypercubes*, using a recently introduced construction known as *double dilation* [29]. The construction is done in two steps: first we define the category DD(fHilb), containing hyper-quantum systems and processes between them, and only in a second moment we introduce quantum and classical systems, using (hyper-) decoherence and working in the Karoubi envelope Split(DD(fHilb)).

The *double-dilation category* DD(fHilb) is defined to be a symmetric monoidal subcategory of CPM(fHilb) with objects—the *density hypercubes*—in the form $DD(H) := \mathcal{H} \otimes \mathcal{H}$, where H is a finite-dimensional Hilbert space and $\mathcal{H} := H^* \otimes H$ is the corresponding doubled system in the CPM category. Even though DD(fHilb) is symmetric monoidal and has its own graphical calculus, in this work we will always use the graphical calculi of CPM(fHilb) and fHilb to talk about density hypercubes. When working in CPM(fHilb), we will use solid black lines for morphisms and calligraphic letters (e.g. $\mathcal{H}$) for objects. When working in fHilb, we will use solid grey lines for morphisms and plain letters (e.g. H) for objects.

The morphisms $DD(H) \to DD(K)$ in DD(fHilb) are the CP maps $\mathcal{H} \otimes \mathcal{H} \to \mathcal{K} \otimes \mathcal{K}$ taking the following form for a doubled CP map F, some auxiliary systems $\mathcal{E}, \mathcal{G}$ and some special commutative †-Frobenius algebra $\circ$ (henceforth known as a *classical structure*) on G in fHilb:

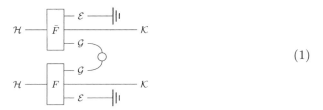

$$(1)$$

In the diagram above, F is a doubled CP map $\mathcal{H} \to \mathcal{G} \otimes \mathcal{K} \otimes \mathcal{E}$ in CPM(fHilb)—i.e. one in the form $F = f^* \otimes f$ for some $f : H \to G \otimes K \otimes E$ in fHilb—and we have used $\bar{F}$ to denote the CP map obtained by inverting the tensor product ordering of inputs and outputs of f (for purely aesthetic reasons). We will always use upper-case letters (e.g. F) to denote doubled CP maps in CPM(fHilb), lower-case letters to denote the corresponding linear maps in fHilb, and we will always write discarding maps explicitly.

Composition in DD(fHilb) is the same as composition of CP maps, while tensor product is only slightly adjusted to take into account the doubled format of our new morphisms:

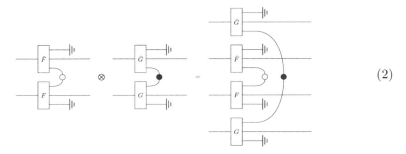

$$(2)$$

Just as was the case for CP maps, maps of density hypercubes can all be obtained as composition of a "doubled" map and one or two "discarding" maps:

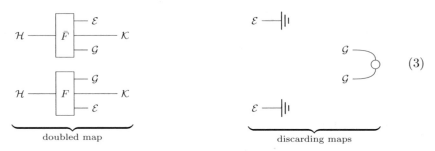

$$(3)$$

We refer to the discarding map obtained by doubling $-\!|\!|_\mathcal{E}$ as the "forest" and to the discarding map obtained from the classical structure $\circ$ as the "bridge". The scalars of DD(fHilb) are exactly the scalars $\mathbb{R}^+$ of CPM(fHilb), and hence the theory of density hypercubes is probabilistic. It is furthermore convex, because the following "tree-on-a-bridge" effects can be used to add-up maps of density hypercubes—analogously to the way ordinary discarding maps $-\!|\!|_\mathcal{H}$ can be used to add-up CP maps in CPM(fHilb)—by expanding them in terms of the orthonormal bases $|\psi_x\rangle_{x \in X}$ associated [8] with the classical structures $\circ$:

$$(4)$$

2.2   Component Symmetries

States in the theory DD(fHilb) take the form of fourth order tensors, an observation which prompted the choice of "density hypercubes" as a name for the theory. If $(|\psi_x\rangle)_{x \in X}$ is a choice of orthonormal basis for some finite-dimensional Hilbert space H, the states on $DD(H) = \mathcal{H} \otimes \mathcal{H}$ in DD(fHilb) can be expanded as follows in fHilb:

$$(5)$$

Recall that density matrices possess a $\mathbb{Z}_2$ symmetry given by self-adjointness. This symmetry can be understood in terms of the following action $\tau : \mathbb{Z}_2 \to$ Aut($\mathbb{C}$) of $\mathbb{Z}_2$ on the complex numbers:

$$\tau(0) := z \mapsto z \qquad\qquad \tau(1) := z \mapsto z^* \tag{6}$$

The components of a density matrix ρ then satisfy the following equation, for every $a \in \mathbb{Z}_2$ (trivial for $a = 0$, self-adjoint for $a = 1$):

$$\tau(a)(\rho_{x_0\,x_1}) = \rho_{x_{(0\oplus a)}\,x_{(1\oplus a)}} \tag{7}$$

Instead of a $\mathbb{Z}_2$ symmetry, density hypercubes possess a $\mathbb{Z}_2 \times \mathbb{Z}_2$ symmetry. This symmetry can be understood in terms of the following action $\tau : \mathbb{Z}_2 \times \mathbb{Z}_2 \to \mathrm{Aut}(\mathbb{C})$ of $\mathbb{Z}_2 \times \mathbb{Z}_2$ on the complex numbers:

$$\begin{aligned} \tau(0,0) &:= z \mapsto z & \tau(0,1) &:= z \mapsto z^* \\ \tau(1,0) &:= z \mapsto z^* & \tau(1,1) &:= z \mapsto z \end{aligned} \tag{8}$$

The components of a density hypercube ρ satisfy the following equation for every $(a, b) \in \mathbb{Z}_2 \times \mathbb{Z}_2$, where by $\oplus$ we have denoted addition in $\mathbb{Z}_2$:

$$\tau(a,b)(\rho_{x_{(0,0)}\,x_{(0,1)}\,x_{(1,0)}\,x_{(1,1)}}) = \rho_{x_{(0\oplus a,0\oplus b)}\,x_{(0\oplus a,1\oplus b)}\,x_{(1\oplus a,0\oplus b)}\,x_{(1\oplus a,1\oplus b)}} \tag{9}$$

We see that the components are related by a trivial symmetry for $a = (0,0)$, by a self-adjoining symmetry for $a = (1,0)$ and $a = (0,1)$, and by a self-transposing symmetry in for $a = (1,1)$. An alternative way to look at this symmetry is observe that states of density hypercubes can all be expressed as certain sums of doubled states in the following form:

$$\begin{array}{c} \boxed{\overline{\Phi}} \!-\!\!-\!\!-\, \mathcal{K} \\ \boxed{\Phi} \!-\!\!-\!\!-\, \mathcal{K} \end{array} \tag{10}$$

For these states, we have the usual self-conjugating $\mathbb{Z}_2$ symmetry of density matrices $\Phi \otimes \overline{\Phi} \mapsto \Phi^* \otimes \overline{\Phi^*}$ as well as an independent self-transposing $\mathbb{Z}_2$ symmetry $\Phi \otimes \overline{\Phi} \mapsto \overline{\Phi} \otimes \overline{\overline{\Phi}}$, which taken together give the same $\mathbb{Z}_2 \times \mathbb{Z}_2$ symmetry described above in terms of components.

In order to visualise the $\mathbb{Z}_2 \times \mathbb{Z}_2$ symmetry action, we divide the components $\rho_{x_{00}x_{01}x_{10}x_{11}}$ of a d-dimensional density hypercube ρ into 15 classes, depending on which indices $x_{00}, x_{01}, x_{10}, x_{11}$ have same/distinct values chosen from the set $\{1, ..., d\}$. We arrange the indices on a square: index 00 is on the top left corner, 10 acts as reflection about the vertical mid-line, 01 acts as reflection about the horizontal mid-line and 11 acts as $180°$ rotation about the centre. We use colours as names for index values in $\{1, ..., d\}$, with distinct colours denoting distinct values.

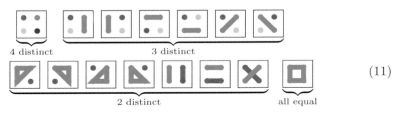

$$\tag{11}$$

For example, the component ρ_{0321} of a 4^+-dimensional system will fall into the 1st class from the left above, the component ρ_{0122} will fall into the 2nd class, the component ρ_{0003} into the 8th class, the component ρ_{0011} into the 12th class and the component ρ_{0000} into the 15th class.

Then we look at the individual orbits of components in each class under the symmetry. Classes with components having orbits of order 4 are shown below: each orbit contributes a single independent complex value to the tensor, i.e. two independent real values, and each component class is annotated by the total number of independent real values contributed in dimension d. Just as we did above, we are using colours to denote values in $\{1, ..., d\}$: the geometric action of $\mathbb{Z}_2 \times \mathbb{Z}_2$ on the coloured vertices/edges of the squares exactly mirrors the algebraic action of $\mathbb{Z}_2 \times \mathbb{Z}_2$ on the components in the different classes.

$$\tag{12}$$

Classes with components having orbits of order 2 and 1 are shown below, each component class annotated by the total number of independent real values contributed in dimension d. Each orbit in the first, second and fourth classes contributes a single independent real value, because each component is stabilised by (at least) one self-adjoining symmetry; each orbit in the third class contributes instead two independent real values, because the components are only stabilised by a self-transposing symmetry.

$$\tag{13}$$

Adding up the contributions from all orbit classes, we see that the states of d-dimensional density hypercubes form a convex cone of real dimension $\frac{1}{2}(d^4 - 3d^3 + 7d^2 - 3d)$ within the $(2d^4)$-dimensional real vector space of complex fourth-order tensors.

2.3   Normalisation and Causality

The "forest" discarding maps $\overline{\overline{\tau}}_{DD(H)} := \text{CPM}(\overline{\tau}_{\mathcal{H}})$ in $DD(\text{fHilb})$ (i.e. the doubled versions of the discarding maps of $\text{CPM}(\text{fHilb})$) form an environment structure [9,11], and we say that a map of density hypercubes is *normalised* if the corresponding CP map is trace preserving (with normalised states as a special case):

$$\text{(14)}$$

Normalised maps of density hypercubes form a sub-SMC of DD(fHilb), which we refer to as the *normalised sub-category*. *Sub-normalised* maps of density hypercubes can be defined analogously by requiring the corresponding CP map to be trace non-increasing: they also form a sub-SMC of DD(fHilb), which we refer to as the *sub-normalised sub-category*.

Despite the presence of several kinds of discarding maps, the following results shows that the sub-normalised sub-category is causal [3], or equivalently that that the normalised sub-category is terminal [6,7].

Proposition 1. *The process theory* DD(fHilb) *is causal, in the following sense: for every object* $DD(H)$, *the only effect* $DD(H) \to \mathbb{R}^+$ *in DD(fHilb) which yields the scalar 1 on all normalised states of* $DD(H)$ *is the "forest" discarding map of density hypercubes* $\text{\reflectbox{F}}_{DD(H)}$.

3   Decoherence and Hyper-decoherence

So far, we have constructed a symmetric monoidal category, which is enriched in convex cones and comes equipped with an environment structure providing a notion of normalization. The final ingredients necessary for the definition of the *categorical probabilistic theory of density hypercubes* is the demonstration that classical systems and quantum systems arise in the Karoubi envelope of DD(fHilb) by choosing some suitable family of decoherence and hyper-decoherence maps.

3.1   Decoherence to Classical Theory

Consider a finite-dimensional Hilbert space H and a classical structure $\circ$ on it, associated with some orthonormal basis $(|\psi_x\rangle)_{x \in X}$. We define the $\circ$-*decoherence map* $\text{dec}_\circ$ on the density hypercube $DD(H)$ to be the following morphism in DD(fHilb):

$$\text{(15)}$$

The $\text{dec}_\circ$ map defined above is idempotent, so it can be used to define classical systems via the Karoubi envelope construction—in the same way as ordinary decoherence maps gives rise to classical systems in quantum theory. It should be noted that decoherence maps defined this way are sub-normalised but not normalised, so that the hyperquantum-to-classical transition in the theory of density hypercubes is not deterministic; we defer further discussion of this point to the next sub-section on hyper-decoherence.

Proposition 2. *Let* $\mathrm{Split}(\mathrm{DD}(\mathrm{fHilb}))$ *be the Karoubi envelope of* DD*(fHilb), and write* $\mathrm{Split}(\mathrm{DD}(\mathrm{fHilb}))_K$ *for the full subcategory of* $\mathrm{Split}(\mathrm{DD}(\mathrm{fHilb}))$ *spanned by objects in the form* $(\mathrm{DD}(H), \mathrm{dec}_{\bigcirc})$. *There is an* $\mathbb{R}^+$*-linear monoidal equivalence of categories between* $\mathrm{Split}(\mathrm{DD}(\mathrm{fHilb}))_K$ *and the probabilistic theory* $\mathbb{R}^+$*-Mat of classical systems. Furthermore, classical stochastic maps correspond to the maps in* $\mathrm{Split}(\mathrm{DD}(\mathrm{fHilb}))_K$ *normalised with respect to the discarding maps* $\overline{\overline{\top}}_{(\mathrm{DD}(H),\mathrm{dec}_{\bigcirc})}$ *defined as* $\overline{\overline{\top}}_{\mathrm{DD}(H)} \circ \mathrm{dec}_{\bigcirc}$ *and which we can write explicitly as follows:*

$$\overline{\overline{\top}}_{(\mathrm{DD}(H),\mathrm{dec}_{\bigcirc})} \quad := \quad \raisebox{-1em}{[diagram]} \quad = \quad \raisebox{-1em}{[diagram]} \tag{16}$$

3.2   Hyper-decoherence to Quantum Theory

We now show that the quantum systems arise in the Karoubi envelope as well, via suitable *hyper-decoherence* maps. Recall that the generic discarding map in the theory of density hypercubes involved two pieces: (the doubled version of) a traditional discarding map from $\mathrm{CPM}(\mathrm{fHilb})$ and a second "tree-on-a-bridge" discarding map derived from a classical structure $\bigcirc$. In the previous sub-section, we saw that the latter is the discarding map of some classical system living in the Karoubi envelope $\mathrm{Split}(\mathrm{DD}(\mathrm{fHilb}))$, and that it can be used to define the "hyper-quantum-to-classical" decoherence maps. In this sub-section, we shall see that this "hyper-quantum-to-classical" decoherence process can be understood in two steps: a "hyper-quantum-to-quantum" hyper-decoherence, followed by the usual "quantum-to-classical" decoherence.

If $\bigcirc$ is a classical structure on a density hypercube $\mathrm{DD}(H)$, we define the $\bigcirc$-*hyper-decoherence map* $\mathrm{hypdec}_{\bigcirc}$ to be the following map of density hypercubes:

$$\mathrm{hypdec}_{\bigcirc} \quad := \quad \raisebox{-1em}{[diagram]} \tag{17}$$

Hyper-decoherence maps are idempotent, and hence we can consider the full subcategory $\mathcal{C}$ of the Karoubi envelope $\mathrm{Split}(\mathrm{DD}(\mathrm{fHilb}))$ spanned by objects in the form $(\mathrm{DD}(H), \mathrm{hypdec}_{\bigcirc})$: doing so allows us to prove that the hyper-decoherence maps defined above truly provide the desired "hyper-quantum-to-quantum" decoherence, as considered by [18, 20].

Proposition 3. *Let* $\mathrm{Split}(\mathrm{DD}(\mathrm{fHilb}))$ *be the Karoubi envelope of* DD*(fHilb), and write* $\mathrm{Split}(\mathrm{DD}(\mathrm{fHilb}))_Q$ *for the full subcategory of* $\mathrm{Split}(\mathrm{DD}(\mathrm{fHilb}))$ *spanned* *by objects in the form* $(DD(H), hypdec_\bigcirc)$. *There is an $\mathbb{R}^+$-linear monoidal equivalence of categories between* $\text{Split}(DD(fHilb))_Q$ *and the probabilistic theory* $\text{CPM}(fHilb)$ *of quantum systems and CP maps between them. Furthermore, trace-preserving CP maps correspond to the maps in* $\text{Split}(DD(fHilb))_Q$ *normalised with respect to the discarding maps* $\overline{\overline{\top}}_{(DD(H), hypdec_\bigcirc)} := \overline{\overline{\top}}_{DD(H)} \circ hypdec_\bigcirc$, *which we can write explicitly as follows:*

$$\overline{\overline{\top}}_{(DD(H), hypdec_\bigcirc)} \quad := \quad \begin{array}{c} \mathcal{H} \\ \mathcal{H} \end{array} \quad = \quad \begin{array}{c} \mathcal{H} \\ \mathcal{H} \end{array} \qquad (18)$$

Taking the double-dilation construction together with the content of Propositions 2 and 3, we come to the following definition of a categorical probabilistic theory [11] of density hypercubes.

Definition 1. *The categorical probabilistic theory of density hypercubes* DH*(fHilb) is defined the be the full sub-SMC of* $\text{Split}(DD(fHilb))$ *spanned by objects in the following form:*

- *the* density hypercubes $(DD(H), id_{DD(H)})$;
- *the* quantum systems $(DD(H), hypdec_\bigcirc)$, *for all classical structures* $\bigcirc$ *on* H;
- *the* classical systems $(DD(H), dec_\bigcirc)$, *for all classical structures* $\bigcirc$ *on* H.

The environment structure for the categorical probabilistic theory is given by the discarding maps $\overline{\overline{\top}}_{DD(H)}$, $\overline{\overline{\top}}_{(DD(H), hypdec_\bigcirc)}$ *and* $\overline{\overline{\top}}_{(DD(H), dec_\bigcirc)}$ *respectively. The classical sub-category for the categorical probabilistic theory is the full sub-SMC spanned by the classical systems.*

The hyper-quantum–to–classical and hyper-quantum–to–quantum decoherence maps of density hypercubes play well together with the quantum–to–classical decoherence map of quantum theory: the decoherence map $dec_\bigcirc$: $(DD(H), id_{DD(H)}) \to (DD(H), dec_\bigcirc)$ of density hypercubes factors, as one would expect, into the hyper-decoherence map $hypdec_\bigcirc$, from $(DD(H), id_{DD(H)})$ to $(DD(H), hypdec_\bigcirc)$, followed by the decoherence map of quantum systems $dec_\bigcirc$, from $(DD(H), hypdec_\bigcirc)$ to $(DD(H), dec_\bigcirc)$. From this, it is clear that the reason why hyper-quantum–to–classical transition was sub-normalised is that the hyper-quantum–to–quantum transition itself is sub-normalised (cf. Appendix B).

The sub-normalisation of hyper-decoherence maps is a sign that the theory of density hypercubes presented here is still partially incomplete, and that some suitable extension will need to be researched in the future. What we know for sure is that the current theory does not satisfy the no-restriction condition on effects, and that an extension in which hyper-decoherence maps are normalised is possible: the additional effect needed by normalisation exists in CPM(fHilb) and is non-negative on all states of DD(fHilb) (cf. Appendix B). In line with the recent no-go theorem of [20], preliminary considerations seem to indicated that the addition of said effect would mean that the theory no longer satisfies purification.

4   Higher Order Interference

In this section, we will show that the theory of density hypercubes displays third- and fourth-order interference effects, broadly inspired by the framework for higher-order interference in GPTs presented by [1, 2, 18]. Because interference has to do with decompositions of the identity map in terms of certain projectors, we begin by introducing a handy graphical notation for keeping track of the various pieces that the identity map is composed of.

The identity map of hyper-quantum systems $id_{\mathrm{DD}(H)} : \mathrm{DD}(H) \to \mathrm{DD}(H)$ takes the following explicit form in fHilb, for any orthonormal basis $(|\psi_x\rangle)_{x \in X}$ of the Hilbert space H:

$$\tag{19}$$

In order to denote the pieces in the decomposition corresponding to specific values $x_{00}, x_{01}, x_{10}, x_{11} \in X$ of the indices, we adopt the following graphical notation, inspired by the $\mathbb{Z}_2 \times \mathbb{Z}_2$ symmetry of the components:

$$\tag{20}$$

In fact, we will adopt the same colour-based notation for index values which we originally introduced in Sect. 2, so that the following is a decomposition piece involving two distinct index values $\{\bullet, \bullet\} \subseteq X$:

$$\tag{21}$$

Using the colour-based notation defined above for its pieces, the identity on a 2-dimensional hyper-quantum system (with $X = \{\bullet, \bullet\}$) would be fully decomposed as follows:

$$id_{\mathbb{C}^2} \;=\; \square + \square + \text{❙❙} + \text{❙❙} + \text{▱} + \text{▱} + \mathsf{X} + \mathsf{X} + \text{◤} + \text{◤} + \text{◥} + \text{◥} + \text{◢} + \text{◢} + \text{◣} + \text{◣} \tag{22}$$

The same notation can be used to graphically decompose projectors corresponding to various subspaces determined by the orthonormal basis $(|\psi_x\rangle)_{x \in X}$. For any non-empty subset $U \subseteq X$, we define the following projector on $\mathrm{DD}(H)$:

$$P_U := \mathrm{DD}\left(\sum_{x \in U} |\psi_x\rangle\langle\psi_x|\right) \tag{23}$$

In particular, the $P_{\{\bullet\}}$ for $\bullet \in X$ are the projectors corresponding to the individual vectors $|\psi_\bullet\rangle$ of the basis, while P_X is the identity $id_{\mathrm{DD}(H)}$. No matter how large X is (with $\#X \geq 2$), the projectors $P_{\{\bullet,\bullet\}}$ corresponding to 2-element subsets $\{\bullet, \bullet\} \subseteq X$ are always decomposed as follows:

$$P_{\{\bullet,\bullet\}} \;=\; \square + \square + \text{❙❙} + \text{❙❙} + \text{▱} + \text{▱} + \mathsf{X} + \mathsf{X} + \text{◤} + \text{◤} + \text{◥} + \text{◥} + \text{◢} + \text{◢} + \text{◣} + \text{◣} \tag{24}$$

The presence of higher order interference in the theory of density hypercubes is really a matter of shapes: when the dimension of $\mathcal{H}$ is at least 3, the identity contains pieces of shapes which do not appear in projectors for 1-element and 2-element subsets. Because of this, in the theory of density hypercubes the probabilities obtained from 1-slit and 2-slit interference experiments will not be enough to explain the probabilities obtained from 3-slit and/or 4-slit experiments; however, the probabilities obtained from 1-slit, 2-slit, 3-slit and 4-slit experiments will always be enough to explain the probabilities obtained in experiments with 5 or more slits.

Below you can see an atlas of all possible shapes that pieces of the identity can take in our graphical notation, together with a note of the smallest dimension that a projector must have to contain pieces of that shape:

$$\tag{25}$$

$$\tag{26}$$

The shape labelled as 1-dimensional only requires a single index value, and hence pieces of that shape appear in all projectors. The shapes labelled as 2-dimensional all require exactly two distinct index values, and hence pieces of those shapes can only appear in projectors for subsets with at least 2 elements. The shapes labelled as 3-dimensional all require exactly three distinct index values, and hence pieces of those shapes can only appear in projectors for subsets with at least 3 elements. Finally, the shape labelled as 4-dimensional requires exactly four index values, and hence pieces of that shape can only appear in projectors for subsets with at least 4 elements.

Thanks to the graphical notation introduced above, we already have a first intuition of why density hypercubes display higher-order interference. However, a rigorous proof requires a complete set-up with states, projectors, measurements and probabilities for a d-slit interference experiment, so that is what we now endeavour to provide.

1. We choose a d-dimensional space $H \cong \mathbb{C}^d$, and we value our tensor indices in the set $X = \{1, ..., d\}$ (the same set that we use to label the d slits).
2. We fix an orthonormal basis $(|x\rangle)_{x \in X}$, and we interpret $|x\rangle$ to be the state in which the particle goes through slit x with certainty.
3. The initial state for the particle is the superposition state in which the particle goes through each slit with the same amplitude. More precisely, it is the pure normalised density hypercube state ρ_+ corresponding to the vector $\frac{1}{\sqrt{d}}|\psi_+\rangle :=$ $\frac{1}{\sqrt{d}}(|1\rangle + ... + |d\rangle)$:

$$\rho_+ := \frac{1}{d^2} \begin{array}{l} \boxed{\overline{\psi_+}} \!-\! \mathcal{H} \\ \boxed{\psi_+} \!-\! \mathcal{H} \end{array} \tag{27}$$

4. The particle goes through some non-empty subset $U \subseteq X$ of slits at random: afterwards, the experimenter knows which subset the particle passed through, but no more information than that is available in the universe.
5. The particle is measured at the screen, and the experimenter estimates the probability $\mathbb{P}[+|U]$ that the particle is still in state ρ_+ after having passed through the given subset U of the slits:

$$\mathbb{P}[+|U] := \frac{1}{d^2} \begin{array}{l} \boxed{\overline{\psi_+}} \!-\! \boxed{\overline{P_U}} \!-\! \boxed{\overline{\psi_+^\dagger}} \\ \boxed{\psi_+} \!-\! \boxed{P_U} \!-\! \boxed{\psi_+^\dagger} \end{array} \frac{1}{d^2} \tag{28}$$

It is immediate to see that the outcome probability $\mathbb{P}[+|U]$ depends solely on the number of different pieces appearing in the decomposition of the projector P_U:

$$\mathbb{P}[+|U] = \frac{1}{d^4} \cdot \text{number of pieces in } P_U \tag{29}$$

To count the number of pieces in P_U, it is convenient to group them by shapes. If U is a subset of size k, standard combinatorial arguments can be used to obtain the number of pieces of each shape appearing in the decomposition (as a convention, we set $\binom{k}{j} = 0$ for $j > k$):

$$\tag{30}$$

By adding up the contributions from pieces of each shape, we get the following closed expression for the outcome probability $\mathbb{P}[+|U]$:

$$\mathbb{P}[+|U] = \frac{1}{d^4}(\#U)^4 \tag{31}$$

For $d \geq 3$ we observe third-order interference, witnessed (by definition) by the following inequality:

$$\mathbb{P}[+|\{1,2,3\}] \neq \sum_{\substack{V \subset \{1,2,3\} \\ \text{s.t. } \#V=2}} \mathbb{P}[+|V] - \sum_{\substack{V \subset \{1,2,3\} \\ \text{s.t. } \#V=1}} \mathbb{P}[+|V] \tag{32}$$

Indeed, the left hand side evaluates to $81/d^4$, while the right hand side evaluates to the following expression (again by standard combinatorial arguments):

$$\frac{1}{d^4}\left[\binom{3}{2}2^4 - \binom{3}{1}1^4\right] = \frac{1}{d^4}45 \neq \frac{1}{d^4}81 \tag{33}$$

The difference between left and right hand sides is $36/d^4$, which is exactly the contribution $\frac{1}{d^4}6 \cdot \binom{3}{3} \cdot 3!$ of the 6 shapes requiring 3 distinct values (appearing in $P_{\{1,2,3\}}$ but not in any of the sub-projectors). For $d \geq 4$ we observe fourth-order interference, witnessed (by definition) by the following inequality:

$$\mathbb{P}[+|\{1,2,3,4\}] \neq \sum_{\substack{V \subset \{1,2,3,4\} \\ \text{s.t. } \#V=3}} \mathbb{P}[+|V] - \sum_{\substack{V \subset \{1,2,3,4\} \\ \text{s.t. } \#V=2}} \mathbb{P}[+|V] + \sum_{\substack{V \subset \{1,2,3,4\} \\ \text{s.t. } \#V=1}} \mathbb{P}[+|V] \tag{34}$$

Indeed, the left hand side evaluates to $256/d^4$, while the right hand side evaluates to the following expression (again by standard combinatorial arguments):

$$\frac{1}{d^4}\left[\binom{4}{3}3^4 - \binom{4}{2}2^4 + \binom{4}{1}1^4\right] = \frac{1}{d^4}232 \neq \frac{1}{d^4}256 \tag{35}$$

The difference between left and right hand sides is $24/d^4$, which is exactly the contribution $\frac{1}{d^4}\binom{4}{4} \cdot 4!$ of the shape requiring 4 distinct values (appearing in $P_{\{1,2,3,4\}}$ but not in any of the sub-projectors).

For $d \geq 5$, however, we observe absence of fifth-order (or higher-order) interference, witnessed (by definition) by the following equality:

$$\mathbb{P}[+|\{1,2,3,4,5\}] = \sum_{\substack{V \subset \{1,2,3,4,5\} \\ \text{s.t. } \#V=4}} \mathbb{P}[+|V] - \sum_{\substack{V \subset \{1,2,3,4,5\} \\ \text{s.t. } \#V=3}} \mathbb{P}[+|V]$$
$$+ \sum_{\substack{V \subset \{1,2,3,4,5\} \\ \text{s.t. } \#V=2}} \mathbb{P}[+|V] - \sum_{\substack{V \subset \{1,2,3,4,5\} \\ \text{s.t. } \#V=1}} \mathbb{P}[+|V] \tag{36}$$

Indeed, the left hand side evaluates to $625/d^4$, and the right hand side yields the same:

$$\frac{1}{d^4}\left[\binom{5}{4}4^4 - \binom{5}{3}3^4 + \binom{5}{2}2^4 - \binom{5}{1}1^4\right] = \frac{1}{d^4}625 \tag{37}$$

5   Conclusions

In this work, we used an iterated CPM construction known as double-dilation to construct a full-fledged probabilistic theory of density hypercubes, possessing hyper-decoherence maps and showing higher-order interference effects. We have defined all the necessary categorical structures. We have gone over the mathematical detail of the (hyper-)decoherence–induced relationship between our new theory, quantum theory and classical theory. We have imported diagrammatic reasoning from the familiar setting of mixed-state quantum theory. We have developed a graphical formalism to study the internal component symmetries of states and processes. Finally, we have shown that the theory displays interference effects of orders up to four, but not of orders five and above.

A number of questions are left open and will be answered as part of future work. Firstly, we endeavour to carry out a more physically-oriented analysis of the theory, including a study of the structure of normalised states and effects and a characterisation of the normalised reversible transformations. Secondly, we need to investigate the physical significance and implications of subnormalisation of the hyper-decoherence maps, and construct a suitable extension of our theory where said maps become normalised. Finally, we intend to look at concrete implementations of certain protocols in our theory, such as those previously studied [16, 18] in the context of higher-order interference.

From a categorical standpoint, we also wish to further understand the specific roles played by double-mixing and double-dilation in our theory. At present, we know that the former is enough for density hypercubes to show higher-order interference and decohere to classical systems, but the latter seems to be necessary for quantum systems to arise by hyper-decoherence. Further investigation will hopefully shed more light on the individual contributions of the two constructions. Finally, we endeavour to investigate the generalisation of our results to higher iterated dilation, and more generally to higher-order CPM constructions [12] (with finite abelian symmetry groups other than the $\mathbb{Z}_2^N$ groups arising from iterated dilation).

Acknowledgements. SG is supported by a grant on Quantum Causal Structures from the John Templeton Foundation. CMS was supported in the writing of this paper by the Engineering and Physical Sciences Research Council (EPSRC) through the doctoral training grant 1652538 and by the Oxford-Google DeepMind graduate scholarship. CMS is currently supported by the Pacific Institute for the Mathematical Sciences (PIMS) and from a Faculty of Science Grand Challenge award at the University of Calgary. This publication was made possible through the support of a grant from the John Templeton Foundation. The opinions expressed in this publication are those of the authors and do not necessarily reflect the views of the John Templeton Foundation.

A   Proofs

Proposition 4. *The process theory* DD(fHilb) *is causal, in the following sense: for every object* $\mathrm{DD}(H)$, *the only effect* $\mathrm{DD}(H) \rightarrow \mathbb{R}^+$ *in* DD*(fHilb) which yields* *the scalar 1 on all normalised states of* $DD(H)$ *is the "forest" discarding map of density hypercubes* $\equiv\mathrel{\!\mid\!\mid}_{DD(H)}$.

Proof. Seen as an effect in CPM(fHilb), any such effect must take the form of a sum $\sum_{x \in X} p_x |a_x\rangle\langle a_x|$, where $p_x \in \mathbb{R}^+$ and $(|a_x\rangle)_{x \in X}$ is an orthonormal basis for $H \otimes H$ which satisfies an additional condition due to the symmetry requirement for effects in DD(fHilb). If we write $\sigma_{H,H}$ for the symmetry isomorphism $H \otimes H \to H \otimes H$ which swaps two copies of H in fHilb, the additional condition on the orthonormal basis implies that for each $x \in X$ there is a unique $y \in X$ such that $\sigma_{H,H}|a_x\rangle = e^{i\theta_x}|a_y\rangle$ and $p_x = p_y$; we define an involutive bijection $s : X \to X$ by setting $s(x)$ to be that unique y. For each $x \in X$, consider the normalised state $\rho_x := \frac{1}{2}(|a_x\rangle\langle a_x| + |a_{s(x)}\rangle\langle a_{s(x)}|)$ in CPM(fHilb), which we can realise in the subcategory DD(fHilb) by considering the classical structure $\bigcirc$ on $\mathbb{C}^2$ corresponding to orthonormal basis $|0\rangle, |1\rangle$ and the vector $|r_x\rangle := \frac{1}{\sqrt[4]{2}}(|a_x\rangle \otimes |0\rangle + |a_{s(x)}\rangle \otimes |1\rangle)$:

$$\rho_x \;=\; \frac{1}{2}\left(\boxed{A_x}\!-\!\mathcal{H} \;+\; \boxed{A_{s(x)}}\!-\!\mathcal{H} \atop \boxed{A_x}\!-\!\mathcal{H} \quad \boxed{A_{s(x)}}\!-\!\mathcal{H} \right) \;=\; \tag{38}$$

Now observe that the requirement that our effect yield 1 on all normalised states implies, in particular, that the following equation must hold:

$$1 \;=\; \quad =\; \frac{1}{2}(p_x + p_{s(x)}) \;=\; p_x \tag{39}$$

As a consequence, our effect is written $\sum_{x \in X} |a_x\rangle\langle a_x|$, which is exactly the "forest" discarding map $\equiv\mathrel{\!\mid\!\mid}_{DD(H)}$ of density hypercubes on $DD(H)$.

Proposition 5. *Let* $\mathrm{Split}(DD(\mathrm{fHilb}))$ *be the Karoubi envelope of* $DD(\mathrm{fHilb})$, *and write* $\mathrm{Split}(DD(\mathrm{fHilb}))_K$ *for the full subcategory of* $\mathrm{Split}(DD(\mathrm{fHilb}))$ *spanned by objects in the form* $(DD(H), \mathrm{dec}_{\bigcirc})$. *There is an* $\mathbb{R}^+$-*linear monoidal equivalence of categories between* $\mathrm{Split}(DD(\mathrm{fHilb}))_K$ *and the probabilistic theory* $\mathbb{R}^+$-*Mat of classical systems. Furthermore, classical stochastic maps correspond to the maps in* $\mathrm{Split}(DD(\mathrm{fHilb}))_K$ *normalised with respect to the discarding maps* $\equiv\mathrel{\!\mid\!\mid}_{(DD(H),\mathrm{dec}_{\bigcirc})}$ *defined as* $\equiv\mathrel{\!\mid\!\mid}_{DD(H)} \circ\, \mathrm{dec}_{\bigcirc}$ *and which we can write explicitly as follows:*

$$\equiv\mathrel{\!\mid\!\mid}_{(DD(H),\mathrm{dec}_{\bigcirc})} \;:=\; \tag{16}$$

Proof. Consider two objects $(\mathrm{DD}(H), \mathrm{dec}_\bullet)$ and $(\mathrm{DD}(K), \mathrm{dec}_\circ)$, where $\bullet$ and $\circ$ are special commutative †-Frobenius algebras associated with orthonormal bases $(|\psi_x\rangle)_{x \in X}$ and $(|\phi_y\rangle)_{y \in Y}$ of H and K respectively. The morphisms from $(\mathrm{DD}(H), \mathrm{dec}_\bullet)$ to $(\mathrm{DD}(K), \mathrm{dec}_\circ)$ in $\mathrm{Split}(\mathrm{DD}(\mathrm{fHilb}))$ are exactly the maps of density hypercubes $\mathrm{DD}(H) \to \mathrm{DD}(K)$ in the following form:

$$\tag{40}$$

We can expand the definition of decoherence maps to see that these morphisms correspond to generic matrices M_{xy} of non-negative real numbers, with matrix composition as sequential composition, Kronecker product as tensor product, and the $\mathbb{R}^+$-linear structure of matrix addition.

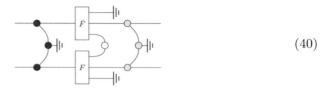

$$\tag{41}$$

The discarding maps obtained by decoherence of the environment structure for $\mathrm{DD}(\mathrm{fHilb})$ yield the usual environment structure for classical systems:

$$\tag{42}$$

Hence $\mathcal{C}_K$ is equivalent to the probabilistic theory $\mathbb{R}^+$-Mat of classical systems.

Proposition 6. *Let* $\mathrm{Split}(\mathrm{DD}(\mathrm{fHilb}))$ *be the Karoubi envelope of* $\mathrm{DD}(\mathrm{fHilb})$, *and write* $\mathrm{Split}(\mathrm{DD}(\mathrm{fHilb}))_Q$ *for the full subcategory of* $\mathrm{Split}(\mathrm{DD}(\mathrm{fHilb}))$ *spanned by objects in the form* $(\mathrm{DD}(H), \mathrm{hypdec}_\circ)$. *There is an* $\mathbb{R}^+$-*linear monoidal equivalence of categories between* $\mathrm{Split}(\mathrm{DD}(\mathrm{fHilb}))_Q$ *and the probabilistic theory* $\mathrm{CPM}(\mathrm{fHilb})$ *of quantum systems and CP maps between them. Furthermore, trace-preserving CP maps correspond to the maps in* $\mathrm{Split}(\mathrm{DD}(\mathrm{fHilb}))_Q$ *normalised with respect to the discarding maps* $\sqcup_{(\mathrm{DD}(H),\mathrm{hypdec}_\circ)} := \sqcup_{\mathrm{DD}(H)} \circ$ $\mathrm{hypdec}_\circ$, *which we can write explicitly as follows:*

$$\tag{18}$$

Proof. We can define an essentially surjective, faithful monoidal functor from $\mathrm{Split}(\mathrm{DD}(\mathrm{fHilb}))$ to the category $\mathrm{CPM}(\mathrm{fHilb})$ of quantum systems and CP maps by setting $(\mathrm{DD}(H), \mathrm{hypdec}_{\bullet}) \mapsto \mathcal{H}$ on objects and doing the following on morphisms:

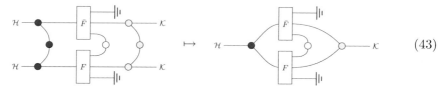

$$\mapsto \qquad (43)$$

In order to show monoidal equivalence we need to show that the functor is also full, i.e. that every CP map can be obtained from a map of $\mathrm{Split}(\mathrm{DD}(\mathrm{fHilb}))$ in this way. Because of compact closure, it is actually enough to show that all states can be obtained this way. Consider a finite-dimensional Hilbert space H and a classical structure $\circ$ on it, and write $(|\psi_x\rangle)_{x \in X}$ for the orthonormal basis of H associated to $\circ$. The most generic mixed quantum state on $\mathcal{H}$ takes the form $\rho = \sum_{y \in Y} p_y |\gamma_y\rangle\langle\gamma_y|$, where $(|\gamma_y\rangle)_{y \in Y}$ is some orthonormal basis of H and $p_y \in \mathbb{R}^+$. Let $\circ$ be the classical structure associated with the orthonormal basis $(|\gamma_y\rangle)_{y \in Y}$, and define the states $|\sqrt[\circ]{\gamma_y}\rangle := \sum_{x \in X} |\psi_x\rangle \sqrt{\langle\psi_x|\gamma_y\rangle}$, where $\sqrt{\langle\psi_x|\gamma_y\rangle} \in \mathbb{C}$ is such that $\sqrt{\langle\psi_x|\gamma_y\rangle}^2 = \langle\psi_x|\gamma_y\rangle \in \mathbb{C}$. If we write $|\phi\rangle := \sum_{yY} \sqrt{p_y} |\sqrt[\circ]{\gamma_y}\rangle \otimes |\gamma_y\rangle$, then the desired state ρ can be obtained as follows:

$$\rho = \sum_{y \in Y} p_y \;\boxed{\Gamma_y}\!\!- \;\; = \;\; \sum_{y \in Y} \frac{\sqrt{p_y}}{\sqrt{p_y}} \;\frac{\boxed{\sqrt[\circ]{\Gamma_y}}}{\boxed{\sqrt[\circ]{\Gamma_y}}} \;\; = \;\; \frac{\boxed{\bar{\Phi}}}{\boxed{\Phi}} \tag{44}$$

Hence the monoidal functor defined above is full, faithful and essentially surjective, i.e. an equivalence of categories. Furthermore, it is $\mathbb{R}^+$-linear and it respects discarding maps.

B Possibility of Extension for the Theory of Density Hypercubes

The theory of density hypercubes presented in this work is fully-fledged[1] but incomplete: as shown by Eq. 18, the hyper-decoherence maps are not normalised (i.e. they are not "deterministic", in the parlance of OPTs/GPTs)

$$\overline{\|}_{(\mathrm{DD}(H), \mathrm{hypdec}_{\bigcirc})} \quad := \qquad \tag{18}$$

---

[1] In the sense that it contains all the features necessary to consistently talk about operational scenarios, such as preparations, measurements, controlled transformations, reversible transformation, test, non-locality scenarios, etc.

When it comes to this work, however, this is not much of a problem: all we need to show is that an extension of our theory can exists in which the "tree-on-a-bridge" effect above can be completed to the discarding map, and our results—both hyper-decoherence to quantum theory and higher-order interference—will automatically apply to any such extension.

Let $(|\psi_x\rangle)_{x \in X}$ be the orthonormal basis associated with the special commutative $\dagger$-Frobenius algebra $\bigcirc$. The effect needed to complete $\text{---}||_{(\mathrm{DD}(H),\mathrm{hypdec}_\bigcirc)}$ to the discarding map $\text{---}||_{\mathrm{DD}(H)}$ is itself an effect in $\mathrm{CPM}(\mathrm{fHilb})$, which can be written explicitly as follows:

$$
\begin{array}{c}
\mathcal{H} \text{---}|| \\
\mathcal{H} \text{---}||
\end{array}
\quad - \quad
\mathcal{H} \text{---}\!\!\!\!\!\begin{array}{c}\mathcal{H}\end{array}\!\!\!\!\!\supset\!\!\!-\!\!||
\quad = \quad
\sum_{\substack{x,y \in X \\ \text{s.t. } x \neq y}}
\begin{array}{c}
\mathcal{H} \text{---}\boxed{\overline{\psi_x^\dagger}} \\
\mathcal{H} \text{---}\boxed{\psi_y^\dagger}
\end{array}
\tag{45}
$$

Because it is an effect in $\mathrm{CPM}(\mathrm{fHilb})$, which has $\mathbb{R}^+$ as its semiring of scalars, it is in particular non-negative on all states in $\mathrm{DD}(\mathrm{fHilb})$, showing that: (i) hyper-decoherence maps are sub-normalised; (ii) our theory does not satisfy the no-restriction condition; (iii) an extension to a theory with normalised hyper-decoherence is possible. This shows that our results on hyper-decoherence have physical significance. Furthermore, let $|1\rangle, ..., |d\rangle$ be an orthonormal basis of $\mathbb{C}^d$, and let $\bigcirc$ correspond to the Fourier basis for the finite abelian group $\mathbb{Z}_d$:

$$
\left(\frac{1}{\sqrt{d}} \sum_{j=1}^{d} e^{i\frac{2\pi}{d}jk} |j\rangle \right)_{k=1,...,d}
\tag{46}
$$

Choosing $k := d$, in particular, shows that the orthonormal basis above contains the state $\frac{1}{\sqrt{d}}|\psi_+\rangle$ used in Sect. 4. Then the effect defined in Eq. 45 also shows that the computation of $\mathbb{P}[+|U]$ in Sect. 4 can be done as part of a bonafide measurement in any such extended theory, and hence that our higher-order interference result has physical significance.

References

1. Barnum, H., Lee, C.M., Scandolo, C.M., Selby, J.H.: Ruling out higher-order interference from purity principles. Entropy **19**(6), 253 (2017). https://doi.org/10.3390/e19060253
2. Barnum, H., Müller, M.P., Ududec, C.: Higher-order interference and single-system postulates characterizing quantum theory. New J. Phys. **16**(12), 123029 (2014). https://doi.org/10.1088/1367-2630/16/12/123029
3. Chiribella, G., D'Ariano, G.M., Perinotti, P.: Probabilistic theories with purification. Phys. Rev. A **81**, 062348 (2010). https://doi.org/10.1103/PhysRevA.81.062348
4. Chiribella, G., Scandolo, C.M.: Entanglement as an axiomatic foundation for statistical mechanics. arXiv:1608.04459 [quant-ph] (2016). http://arxiv.org/abs/1608.04459

5. Chiribella, G., Scandolo, C.M.: Microcanonical thermodynamics in general physical theories. New J. Phys. **19**(12), 123043 (2017). https://doi.org/10.1088/1367-2630/aa91c7
6. Coecke, B.: Terminality implies no-signalling... and much more than that. New Gener. Comput. **34**(1–2), 69–85 (2016). https://doi.org/10.1007/s00354-016-0201-6
7. Coecke, B., Lal, R.: Causal categories: relativistically interacting processes. Found. Phys. **43**(4), 458–501 (2013). https://doi.org/10.1007/s10701-012-9646-8
8. Coecke, B., Pavlovic, D., Vicary, J.: A new description of orthogonal bases. Math. Struct. Comput. Sci. **23**(3), 555–567 (2013)
9. Coecke, B., Perdrix, S.: Environment and classical channels in categorical quantum mechanics. In: Dawar, A., Veith, H. (eds.) CSL 2010. LNCS, vol. 6247, pp. 230–244. Springer, Heidelberg (2010). https://doi.org/10.1007/978-3-642-15205-4_20
10. Dakić, B., Paterek, T., Brukner, Č.: Density cubes and higher-order interference theories. New J. Phys. **16**(2), 023028 (2014). https://doi.org/10.1088/1367-2630/16/2/023028
11. Gogioso, S., Scandolo, C.M.: Categorical probabilistic theories. In: Coecke, B., Kissinger, A. (eds.) Proceedings 14th International Conference on Quantum Physics and Logic, Nijmegen, The Netherlands, 3–7 July 2017. Electronic Proceedings in Theoretical Computer Science, vol. 266, pp. 367–385. Open Publishing Association (2018). https://doi.org/10.4204/EPTCS.266.23
12. Gogioso, S.: Higher-order CPM constructions. Electron. Proc. Theor. Comput. Sci. **270**, 145–162 (2019). https://doi.org/10.4204/EPTCS.287.8
13. Jin, F., et al.: Experimental test of born's rule by inspecting third-order quantum interference on a single spin in solids. Phys. Rev. A **95**, 012107 (2017). https://doi.org/10.1103/PhysRevA.95.012107
14. Kauten, T., Keil, R., Kaufmann, T., Pressl, B., Brukner, Č., Weihs, G.: Obtaining tight bounds on higher-order interferences with a 5-path interferometer. New J. Phys. **19**(3), 033017 (2017). https://doi.org/10.1088/1367-2630/aa5d98
15. Krumm, M., Barnum, H., Barrett, J., Müller, M.P.: Thermodynamics and the structure of quantum theory. New J. Phys. **19**(4), 043025 (2017). https://doi.org/10.1088/1367-2630/aa68ef
16. Lee, C.M., Selby, J.H.: Deriving Grover's lower bound from simple physical principles. New J. Phys. **18**(9), 093047 (2016). https://doi.org/10.1088/1367-2630/18/9/093047
17. Lee, C.M., Selby, J.H.: Generalised phase kick-back: the structure of computational algorithms from physical principles. New J. Phys. **18**(3), 033023 (2016). https://doi.org/10.1088/1367-2630/18/3/033023
18. Lee, C.M., Selby, J.H.: Higher-order interference in extensions of quantum theory. Found. Phys. **47**(1), 89–112 (2017). https://doi.org/10.1007/s10701-016-0045-4
19. Lee, C.M., Selby, J.H., Barnum, H.: Oracles and query lower bounds in generalised probabilistic theories. arXiv:1704.05043 [quant-ph] (2017). https://arxiv.org/abs/1704.05043
20. Lee, C.M., Selby, J.H.: A no-go theorem for theories that decohere to quantum mechanics. Proc. Roy. Soc. A: Math. Phys. Eng. Sci. **474**(2214), 20170732 (2018). https://doi.org/10.1098/rspa.2017.0732
21. Niestegge, G.: Three-slit experiments and quantum nonlocality. Found. Phys. **43**(6), 805–812 (2013). https://doi.org/10.1007/s10701-013-9719-3
22. Park, D.K., Moussa, O., Laflamme, R.: Three path interference using nuclear magnetic resonance: a test of the consistency of Born's rule. New J. Phys. **14**(11), 113025 (2012). https://doi.org/10.1088/1367-2630/14/11/113025

23. Sinha, A., Vijay, A.H., Sinha, U.: On the superposition principle in interference experiments. Sci. Rep. **5**, 10304 (2015). https://doi.org/10.1038/srep10304
24. Sinha, U., Couteau, C., Jennewein, T., Laflamme, R., Weihs, G.: Ruling out multi-order interference in quantum mechanics. Science **329**(5990), 418–421 (2010). https://doi.org/10.1126/science.1190545
25. Sorkin, R.D.: Quantum mechanics as quantum measure theory. Mod. Phys. Lett. A **9**(33), 3119–3127 (1994). https://doi.org/10.1142/S021773239400294X
26. Sorkin, R.D.: Quantum Measure Theory and its Interpretation. In: Quantum Classical Correspondence: The 4th Drexel Symposium on Quantum Nonintegrability, pp. 229–251. International Press, Boston (1997)
27. Ududec, C.: Perspectives on the formalism of quantum theory. Ph.D. thesis, University of Waterloo (2012)
28. Ududec, C., Barnum, H., Emerson, J.: Three slit experiments and the structure of quantum theory. Found. Phys. **41**(3), 396–405 (2011). https://doi.org/10.1007/s10701-010-9429-z
29. Zwart, M., Coecke, B.: Double dilation ≠ double mixing (extended abstract). In: Coecke, B., Kissinger, A. (eds.) Proceedings 14th International Conference on Quantum Physics and Logic, Nijmegen, The Netherlands, 3–7 July 2017. Electronic Proceedings in Theoretical Computer Science, vol. 266, pp. 133–146. Open Publishing Association (2018). https://doi.org/10.4204/EPTCS.266.9
30. Życzkowski, K.: Quartic quantum theory: an extension of the standard quantum mechanics. J. Phys. A **41**(35), 355302 (2008). https://doi.org/10.1088/1751-8113/41/35/355302

Information Retrieval

Investigating Non-classical Correlations Between Decision Fused Multi-modal Documents

Dimitris Gkoumas[1]([✉])[iD], Sagar Uprety[1][iD], and Dawei Song[1,2][iD]

[1] The Open University, Milton Keynes, UK
{dimitris.gkoumas,sagar.uprety,dawei.song}@open.ac.uk
[2] Beijing Institute of Technology, Beijing, China

Abstract. Correlation has been widely used to facilitate various information retrieval methods such as query expansion, relevance feedback, document clustering, and multi-modal fusion. Especially, correlation and independence are important issues when fusing different modalities that influence a multi-modal information retrieval process. The basic idea of correlation is that an observable can help predict or enhance another observable. In quantum mechanics, quantum correlation, called entanglement, is a sort of correlation between the observables measured in atomic-size particles when these particles are not necessarily collected in ensembles. In this paper, we examine a multimodal fusion scenario that might be similar to that encountered in physics by firstly measuring two observables (i.e., text-based relevance and image-based relevance) of a multi-modal document without counting on an ensemble of multi-modal documents already labeled in terms of these two variables. Then, we investigate the existence of non-classical correlations between pairs of multi-modal documents. Despite there are some basic differences between entanglement and classical correlation encountered in the macroscopic world, we investigate the existence of this kind of non-classical correlation through the Bell inequality violation. Here, we experimentally test several novel association methods in a small-scale experiment. However, in the current experiment we did not find any violation of the Bell inequality. Finally, we present a series of interesting discussions, which may provide theoretical and empirical insights and inspirations for future development of this direction.

Keywords: Multi-modal information retrieval · Non-classical correlations · Decision fused multi-modal documents · CHSH inequality

1 Introduction

Nowadays, the Web surrounding us often involves multiple modalities - we read texts, watch images and videos, and listen to sounds. In general terms, modality refers to a certain type of information and/or the representation format in which information is stored. A research problem is characterized as multi-modal

© Springer Nature Switzerland AG 2019
B. Coecke and A. Lambert-Mogiliansky (Eds.): QI 2018, LNCS 11690, pp. 163–176, 2019.
https://doi.org/10.1007/978-3-030-35895-2_11

when it includes multiple such modalities. Integrating unimodal representations from various input modalities and combining them into a compact multi-modal representation, called multi-modal fusion, offers a possibility of understanding in-depth real world problems. For instance, in information retrieval, suppose a user types in a text query to retrieve multi-modal documents consisting of an image and a caption as shown in Fig. 1. One can notice that the query term "plane" can be matched in both textual and visual modalities of the given multimodal document. However, the query term "London" can be matched only in its textual modality, while the term "sunset" only in its visual modality. This implies that only when the text and image modalities are fused, we get the benefit of complementary information, in turn increasing the precision of information retrieval.

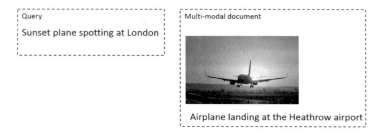

Fig. 1. Example of multi-modal information retrieval

The main challenge of multi-modal fusion is to capture inter-dependencies and complementary presence in heterogeneous data originating from multiple modalities. In the literature, two main approaches to the fusion process have been proposed: (a) *feature level* or *early fusion* and (b) *decision level* or *late fusion* [4]. Early fusion involves the integration of multiple sources of raw or preprocessed data to be fed into a model, which finally makes an inference as illustrated in Fig. 2. In contrast, late fusion refers to the aggregation of decisions from multiple classifiers, each trained on separate modalities as shown in Fig. 3.

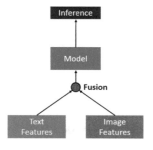

Fig. 2. Early fusion

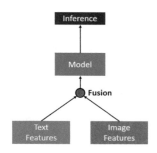

Fig. 3. Late fusion

There are distinctive issues that influence the multi-modal fusion process. Correlation between different modalities is one of them. Correlation can be perceived either in low-level features, e.g., raw data, or high-level features that are obtained on different classifiers, e.g., semantic concepts [4]. In both cases, correlation informs us how to fuse different modalities. In the early fusion, we fuse multi-modal information either by projecting all of the modalities to the same space (Fig. 4(c)), called joint representations, or by learning separate representations for each modality but coordinate them through a similarity measure (Fig. 4(b)) [5]. In both approaches, the construction of the multi-modal spaces is based on correlations among different modalities. The late fusion process can be rule-based, e.g., by linear weighted fusion and majority voting rules, or based on classification-based methods, e.g., support vector machines [4]. In many cases, the correlation among different modalities provides additional cues that are very useful for aggregating decisions either by following a rule-based approach or a classification-based approach. In addition, the absence of correlation may equally provide valuable insight with respect to a particular scenario or context.

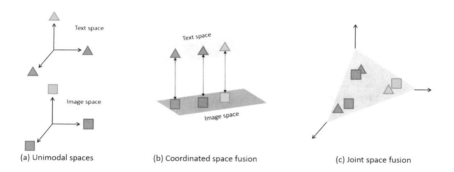

Fig. 4. Construction of multi-modal spaces

There are various statistical and probabilistic forms of correlation that have been utilized by researchers, being causal or not. Since our experiment focuses on late fusion only, we briefly report the most important methods for computing correlations between decisions from multiple modalities. Specifically, decision level correlation has been exploited in the form of causal link analysis, causal strength, and agreement coefficient [4]. In all cases, the basic idea of correlation is that a modality can help predict or enhance another modality.

In quantum mechanics, correlation has been also an important topic. In quantum mechanical framework, uncertainty may occur not only when the elements are collected in ensemble but also when each of them is in a superposed state. In quantum theory, making an observation on one part of a system *instantaneously* could affect the state in another part of a system, even if the respective systems are separated by space-like distances. Such a quantum correlation presents some peculiarities which led to the notion of entanglement. Entanglement is a sort of correlation between observables measured in atomic-size particles, such as photons, when these particles are not necessarily collected in ensembles.

Despite entanglement being a kind of correlation, there are some basic differences between entanglement and the classical correlation encountered in the macroscopic world. A classical correlation is a statistical relationship, causal or not, between two random variables. In entanglement, besides correlation, cause exists as well since the correlation does not depend on an underlying value attached to the particles. Instead, it depends on what is measured on either side. This non-classical property of quantum entanglement motivates us to investigate non-classical correlations between multi-modal decisions as shown in Fig. 5. At first, we calculate the probability of relevance for each document, with respect to both text-based and image-based modality concerning a multimodal query as shown in Fig. 5. Then, we check for any violation of Bell's inequalities based on the estimated relevance probabilities for each possible pair of decision fused multimodal documents in a dataset. Our assumption is that if a pair of decision fused multi-modal documents is *entangled*, then knowing that a document is relevant concerning the text-based representation for a query, then we can *simultaneously* predict with certainty the relevance of the other document concerning the text-based and image-based representation for the same query.

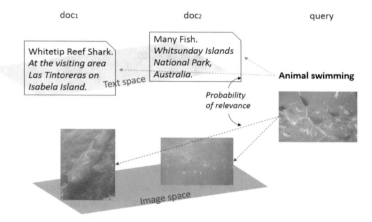

Fig. 5. Investigation of non-classical correlations between decision fused multi-modal documents

The rest of the paper is organized as follows. Section 2 presents a brief review of related work. In Sect. 3 we provide a foundation in quantum entanglement and Bell inequality, while in Sect. 4 we explain some basic concepts of geometry in information retrieval and then formalize the proposed model. Section 5 reports all the experiment settings. In Sect. 6 we report and discuss the results. Finally, Sect. 7 concludes the paper.

2   Related Work

A composite system being entangled cannot be validly decomposed and modeled as separate subsystems. The quantum theory provides formal tools to model interacting systems as non-decomposable in macroscopic world as well. The phenomenon of quantum entanglement has been investigated in semantic spaces making use of Hyperspace Analogue to Language (HAL) model [13,14]. Hou et al. considered high order entanglements that cannot be reduced to the compositional effect of lower-order ones, as an indicator of high-level semantic entities. Melucci proposes quantum-like entanglement for modeling the interaction between a user and a document as a composite system [15].

The non-compositionality of entangled systems opened also the door to developing quantum-like models of cognitive phenomena which are not decompositional in nature. Concept combinations have been widely modeled as composite systems [1,2,6,7,22]. The state of the composite system between two words can be modeled by taking the tensor product of the states of the individual words respectively. If the concept combination is factorizable, then the concept is compositional in the sense it can be expressed as a product of states corresponding to the separate words. A concept that is not factorizable cannot be expressed by either the first or the second word individually, and is deemed *non-compositional*, and termed *entangled* [7].

Quantum theory provides a well-developed set of analytical tools that can be used to determine whether the state of a system of interest can be validly decomposed into separate sub-systems. A possible way to test the non-compositional state of a composite system is the violation of Bell's inequalities. For instance, having calculated the expectation values of variables associated with an experiment, we can fit the Clauser-Horne-Shimony-Holt (CHSH) version of Bell's inequality [9]. If the CHSH inequality is greater than 2, then the Bell inequality is violated. It has been empirically found that the maximal possible violation in quantum theory is $2\sqrt{2} \approx 2.8284$ [8]. This means that each violation being close to the maximal value is very significant. In addition to the CHSH inequality, Bruza et al. [7] propose Clauser-Horne inequalities to analyse the decomposability of quantum systems. The Schmidt decomposition is another way for detecting entanglement in bipartite systems [17]. According to the theorem, after decomposition, each pure state of the tensor product space can be expressed as the product of subsystem orthonormal bases and non-negative real coefficients. The square sum of the coefficients is equal to 1. The number of non-zero coefficients is called Schmidt number. If it equals 1, then the composite state is the product state. If it greater than 1, then the composite state is non-compositional.

So far, researchers have used joint probabilities in cognitive science for calculating expectation values assuming that the outcomes of observables are dependent. Additionally, probabilities can be calculated via trace formula in Gleason's theory [11]. In a similar way, expectation value of two random variables is defined the product of traces [15]. Finally, probabilities could be re-expressed as function of an angle θ, where θ is defined as a difference in phase between two random

observables, once we view the relationship between them as a geometrical relationship [15].

3   Quantum Entanglement and Bell Inequality

Let us suppose that we have a system of two qubits expressed in a Bit basis
$\{0, 1\}$, such that the first qubit is in a state $a_0|0\rangle + a_1|1\rangle$ and the second one in
a state $b_0|0\rangle + b_1|1\rangle$. The state of the two qubits together as a composite system
is a superposition of four classical probabilities resulting in

$$|\phi\rangle = a_0b_0|00\rangle + a_0b_1|01\rangle + a_1b_0|10\rangle + a_1b_1|11\rangle. \tag{1}$$

Let us now assume that the composite system is in an entangled state given by
the following Bell state

$$|\psi\rangle = \frac{1}{\sqrt{2}}|00\rangle + \frac{1}{\sqrt{2}}|11\rangle. \tag{2}$$

When we measure the composite system, the probability of the system to collapse
either to the state $|00\rangle$ or to the state $|11\rangle$ is equal to 0.5. However, after a
measurement, the system is not in an entangled state anymore. For instance,
once we measure the state $|00\rangle$, the new state of the system results in

$$|\psi\rangle = |00\rangle. \tag{3}$$

Let us now assume that we measure the state $|0\rangle$ of the first qubit (Eq. (2)).
Then the probability for the first qubit to collapse to the state $|0\rangle$ again equals
0.5. However, after the measurement, the probability of the second qubit to be
in the state $|0\rangle$ currently equals 1. Let us suppose that we change the Bit basis
to a Sign basis $\{-, +\}$. According to the rotation invariance [19], the Bell state
in the Sign basis is again an equal superposition of the state $|--\rangle$ and the state
$|++\rangle$ such that

$$|\psi\rangle = \frac{1}{\sqrt{2}}|00\rangle + \frac{1}{\sqrt{2}}|11\rangle \tag{4}$$
$$= \frac{1}{\sqrt{2}}|--\rangle + \frac{1}{\sqrt{2}}|++\rangle.$$

Suppose now that we want to measure the probability of the second qubit to be
in the state $|-\rangle$ according to the Sign basis, given that we have already measured
the probability of the first qubit to be in state $|0\rangle$ concerning the Bit basis. Once
we measure the first qubit, the probability of the second qubit to be in the same
state $|0\rangle$ is equal to 1. If θ is the angle between the Bit and Sign basis, then
according to the Pythagorean theorem, the probability of the second qubit to
be in the state $|-\rangle$ equals $\cos^2 \theta$.

   In quantum mechanics, the criteria used to test entanglement are given by
Bell's inequalities. A possible way to proceed is to define four observables. Each
observable has binary values ± 1 thus give two mutually exclusive outcomes. For

instance, a photon can be detected by '+' or '−' channel (see Fig. 9). Let us denote as A_1, B_1 the observables describing the first system, and A_2, B_2 the observables of the second one. If a composite system is separable, the following CHSH inequality holds:

$$|\langle A_1 A_2 \rangle + \langle A_2 B_1 \rangle + \langle A_1 B_2 \rangle - \langle B_1 B_2 \rangle| \leq 2, \tag{5}$$

where $\langle \rangle$ denotes the expectation value between two observables. The calculation of expectation values will be articulated in Sect. 4. The violation of (5) is a sign of entanglement. A Bell inequality violation implies that at least one of the assumptions of *local-realism* made in the proof of (5) must be incorrect [16]. This points to the conclusion that either or both of locality - an object is only directly influenced by its immediate surroundings - and realism - an object has definite values - must be rejected as a property of the composite systems violating CHSH inequality.

4   Non-classical Correlations in Decision Fused Multimodal Documents

Before the late fusion process, there exists a probability $p(R|T)$ for a multimodal document D_M to be relevant to a multimodal information need concerning the textual information. Similarly, the probability for the same document not to be relevant is denoted as $p(\overline{R}|T)$, which is equal to $1-p(R|T)$. Let us consider a real-valued two dimensional Hilbert Space for the relevance of the D_M concerning the textual information (Fig. 6). In Fig. 6 the vector R_t stands for the relevance of the document concerning the text-based modality. On the other hand, the $\overline{R_t}$ represents the non-relevance with respect to the same text-based information need and is orthogonal to R_t.

The text-based relevance of a document can be modeled as a vector in the Hilbert Space, which unifies the logical, probabilistic and vector space based approaches to IR [21]. This vector is a superposition of relevance and non-relevance vectors with respect to the text-based modality and is represented as:

$$|D_M\rangle = a|R_t\rangle + a'|\overline{R_t}\rangle, \tag{6}$$

where $|a|^2 + |a'|^2 = 1$. The coefficients a and a' are captured by taking the projection of $|D_M\rangle$ onto the relevance and non-relevance vectors respectively (Fig. 6) by taking their inner products. According to the Born rule, $p(R|T)$ equals to the square of the inner product $|\langle R_t|D_M\rangle|^2$ and likewise, $p(\overline{R}|T)$ equals to $|\langle \overline{R_t}|D_M\rangle|^2$.

In a similar way, we denote as $p(R|I)$ the probability for a multimodal document D_M to be relevant concerning the image-based information need, and $p(\overline{R}|I)$ the probability to be irrelevant respectively (Fig. 7). The relevance of a document with respect to the image-based modality can be in a similar manner modeled as:

$$|D_M\rangle = b|R_i\rangle + b'|\overline{R_i}\rangle \tag{7}$$

In this case, $p(R|I)$ is computed as the square of the inner product $|\langle R_i|D_M\rangle|^2$. Likewise, $p(\overline{R}|I)$ equals to $|\langle \overline{R_i}|D_M\rangle|^2$.

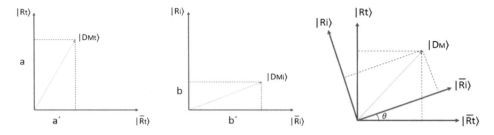

Fig. 6. Text-based relevance in two-dimensional Hilbert Space

Fig. 7. Image-based relevance in two-dimensional Hilbert Space

Fig. 8. Hilbert Space after fusion having a text and image basis

After the late fusion process, the document can be judged based on both text-based and image-based modalities. Such a phenomenon can be modeled in the same Hilbert Space by having a different basis for each modality, as presented in Fig. 8. The document D_M is represented as a unit vector and its representation is expressed with respect to the bases $T = \{|R_t\rangle, |\overline{R_t}\rangle\}$ and $I = \{|R_i\rangle, |\overline{R_i}\rangle\}$ fusing at the end the local decisions. Each basis models context with respect to a given modality.

The rest of the experimental setup is analogous to that one for investigating quantum entanglement in photons [3]. Figure 9 shows the experimental setup for the violation of Bell's inequalities. The source S produces a pair of photons, sent in opposite directions. Each photon encounters a two-channel polariser whose orientation can be set by the experimenter. Coincidences (simultaneous detections) are recorded, the results being categorised as $++, +-, -+, or --$ and corresponding counts accumulated by the coincidence monitor CM.

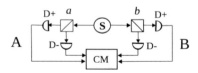

Fig. 9. Schematic of a "two-channel" Bell test

Now let us consider Fig. 10, which depicts two multimodal documents, D_{M1} and D_{M2} respectively. As afore-mentioned, the documents D_{M1} and D_{M2} can be expressed with either the text-based basis or image-based basis being in

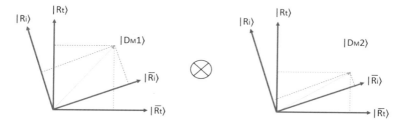

Fig. 10. The interaction between two documents is modeled as a composite system

a superposition of relevance and non-relevance states. In quantum theory, the interaction between D_{M1} and D_{M2} can be modeled as a composited system by using the tensor product of the document Hilbert Spaces. The state of the composite system $D_{M1} \otimes D_{M2}$ can be obtained by taking the tensor product of the relevance and non-relevance states. Concerning the text-based modality, the state of the composite system is defined as follows:

$$
\begin{aligned}
|D_{M1}\rangle \otimes |D_{M2}\rangle &= (a_1|R_t\rangle + a_1'|\overline{R_t}\rangle) \otimes (a_2|R_t\rangle + a_2'|\overline{R_t}\rangle) \\
&= a_1 a_2 |R_t R_t\rangle + a_1 a_2' |R_t \overline{R_t}\rangle + a_1' a_2 |\overline{R_t} R_t\rangle + a_1' a_2' |\overline{R_t R_t}\rangle,
\end{aligned}
\tag{8}
$$

where $|a_1 a_2|^2 + |a_1 a_2'|^2 + |a_1' a_2|^2 + |a_1' a_2'|^2 = 1$. In a similar way, if we define the image-based basis as a standard basis then the state of the composite system $D_{M1} \otimes D_{M2}$ concerning the image-based modality can be expressed as follows:

$$
\begin{aligned}
|D_{M1}\rangle \otimes |D_{M2}\rangle &= (b_1|R_i\rangle + b_1'|\overline{R_i}\rangle) \otimes (b_2|R_i\rangle + b_2'|\overline{R_i}\rangle) \\
&= b_1 b_2 |R_i R_i\rangle + b_1 b'2 |R_i \overline{R_i}\rangle + b_1' b_2 |\overline{R_i} R_i\rangle + b_1' b_2' |\overline{R_i R_i}\rangle
\end{aligned}
\tag{9}
$$

In Eq. (8), the first and second terms reveal that when the text-based content of the D_{M1} is relevant, then we cannot be sure about the relevance of the text-based content of the other document. Similarly, according to the third and fourth term in Eq. (8), when the text-based content of the D_{M1} is non-relevant, then the other document is in a superposition of relevance and non-relevance states with respect to the text-based modality. The same is observed when we consider the image-based basis as a standard basis.

If the state of the text-based (Eq. (8)) or image-based (Eq. (9)) composite system is factorizable, then the system is compositional in the sense it can be expressed as a product of states corresponding to the separate subsystems. A composite quantum system that is not factorizable is deemed *non-compositional* and termed *entangled* [7]. In the last case, if we consider the text-based representation as a standard basis, then we can define two Bell states, either the state

$$
|D_M\rangle = a_1 a_2 |R_t R_t\rangle + a_1' a_2' |\overline{R_t R_t}\rangle,
\tag{10}
$$

or

$$
|D_M\rangle = a_1 a_2' |R_t \overline{R_t}\rangle + a_1' a_2 |\overline{R_t} R_t\rangle.
\tag{11}
$$

Concerning the Eq. (10), the probability for both documents to be relevant (i.e., the state $|R_t R_t\rangle$) regarding the text-based modality equals $|a_1 a_2|^2$. If we measure only the probability of the first document to be relevant concerning the text-based modality results again in $|a_1 a_2|^2$. Then after the measurement, the probability for the second document to be relevant is equal to 1. Consequently, we can *simultaneously* predict the probability of the second document to be relevant concerning the image-based modality, which is equal to $cos^2\theta$, where θ is the angle between the image-based and text-based basis (Fig. 10). Similar outcomes result once we measure the probability for both documents to be irrelevant (i.e., the state $|\overline{R_t R_t}\rangle$ in Eq. (10)), one relevant and the other irrelevant (i.e., the state $|R_t \overline{R_t}\rangle$ in Eq. (11))), or one irrelevant and the other relevant (i.e., the state $|\overline{R_t} R_t\rangle$ in Eq. (11)).

In Sect. 3, we have described the CHSH inequality defining four observables, where each observable has two binary values ± 1 thus gives two mutually exclusive outcomes. In a similar manner, in our case, for the document D_{M1}, we have variables R_{t1} and R_{i1}, which take values $1, -1$, where $R_{t1} = 1$ corresponds to the basis vector $|R_{t1}\rangle$ and $R_{t1} = -1$ corresponds to its orthogonal basis vector $|\overline{R_{t1}}\rangle$. Similarly, $R_{i1} = 1$ corresponds to the basis vector $|R_{i1}\rangle$ and $R_{i1} = -1$ corresponds to its orthogonal basis vector $|\overline{R_{i1}}\rangle$. For the document D_{M2}, we have variables R_{t2} and R_{i2} which take values $1, -1$, where $R_{t2} = 1$ corresponds to the basis vector $|R_{t2}\rangle$ and $R_{t2} = -1$ corresponds to its orthogonal basis vector $|\overline{R_{t2}}\rangle$. Similarly, $R_{i2} = 1$ corresponds to the basis vector $|R_{i2}\rangle$ and $R_{i2} = -1$ corresponds to its orthogonal basis vector $|\overline{R_{i2}}\rangle$. Then Eq. (5) results in

$$|\langle R_{t1} R_{t2}\rangle + \langle R_{t2} R_{i1}\rangle + \langle R_{t1} R_{i2}\rangle - \langle R_{i1} R_{i2}\rangle| \leq 2, \tag{12}$$

where

$$\langle R_{t1} R_{t2}\rangle = ((+1)p(R_{t1}) + (-1)p(\overline{R_{t1}})) * ((+1)p(R_{t2}) + (-1)p(\overline{R_{t2}}))$$
$$= p(R_{t1})p(R_{t2}) - p(R_{t1})p(\overline{R_{t2}}) - p(\overline{R_{t1}})p(R_{t2}) + p(\overline{R_{t1}})p(\overline{R_{t2}}),$$

$$\langle R_{t2} R_{i1}\rangle = ((+1)p(R_{t2}) + (-1)p(\overline{R_{t2}})) * ((+1)p(R_{i1}) + (-1)p(\overline{R_{i1}}))$$
$$= p(R_{t2})p(R_{i1}) - p(R_{t2})p(\overline{R_{i1}}) - p(\overline{R_{t2}})p(R_{i1}) + p(\overline{R_{t2}})p(\overline{R_{i1}}),$$

$$\langle R_{t1} R_{i2}\rangle = ((+1)p(R_{t1}) + (-1)p(\overline{R_{t1}})) * ((+1)p(R_{i2}) + (-1)p(\overline{R_{i2}}))$$
$$= p(R_{t1})p(R_{i2}) - p(R_{t1})p(\overline{R_{i2}}) - p(\overline{R_{t1}})p(R_{i2}) + p(\overline{R_{t1}})p(\overline{R_{i2}}),$$

$$\langle R_{i1} R_{i2}\rangle = ((+1)p(R_{i1}) + (-1)p(\overline{R_{i1}})) * ((+1)p(R_{i2}) + (-1)p(\overline{R_{i2}}))$$
$$= p(R_{i1})p(R_{i2}) - p(R_{i1})p(\overline{R_{i2}}) - p(\overline{R_{i1}})p(R_{i2}) + p(\overline{R_{i1}})p(\overline{R_{i2}}).$$

The above products of probabilities are defined as joint probabilities between two independent outcomes. The violation of Eq. (12) is a sign of entanglement, and the pair of documents may result in one of the aforementioned Bell states (Eqs. (10), (11)) as have been described above.

5   Experiment Settings

5.1   Dataset

The proposed model is tested on the ImageCLEF2007 data collection [12], the purpose of which is to investigate the effectiveness of combining image and text for retrieval tasks. Out of 60 test queries we randomly picked up 30 ones, together with the ground truth data. Each query describing user information need consists of three sample images and a text description, whereas each document consists of an image and a text description. For every query, we created a subset of 300 relevant and irrelevant documents, which includes firstly all the relevant documents for the query, and the rest being irrelevant documents. The dataset is used for investigating both the Bell states (Eqs. (10) and (11)). The number of relevant documents per query ranges from 11 to 98.

5.2   Image and Text Representations-Mono-modal Baselines

The late fusion process is based on mono-modal retrieval scores. For the visual information, feature extraction consists of using the representations learned by the VGG16 model [18], with weights pre-trained on ImageNet to extract features from images, resulting in a feature vector of 2048 floating values for each image. After feature vector extractions, we compute the similarity scores between a submitted visual query and images in the dataset based on Cosine function. For textual information, a query expansion approach has been applied extending the query with the ten most frequent terms according to the ground truth text-based documents. This indeed corresponds to a simulated explicit relevance feedback scenario. Then, the TF-IDF vector representation is used for calculating the text-based Cosine similarity between the a query and text documents. Cosine similarity is particularly used in positive space, where the Cosine similarity score is bounded in [0,1]. In our case, we make use of Cosine similarity score for approximating the probability of relevance.

5.3   Experimental Procedure

At the first step, for both text-based and image-based modalities, the Cosine function is employed to approximate the probability of relevance according to a multi-modal query (Fig. 5). Then, we create pairs of relevant documents. In the next step, expectation values are computed based on probabilities of relevance according to the process being described in Sect. 4. The probability for a document to be relevant concerning a modality is equal to the result of Cosine function. Consequently, the probability for a document to be irrelevant concerning the same modality equals 1 minus the result of Cosine function. Then, we fit the CHSH inequality with the calculated expectation values and check for any existence of violation. For each query, we calculate in total the percentage of documents show a violation of the CHSH inequality. At the end of the experiment, we calculate the percentage of queries showing violation.

6   Results and Discussion

The experiment results are out of our expectations since we did not observe any violation of Bell's inequality. This implies that in the context of our experimental setting non-classical correlations between pairs of documents may not exist, but also that the hypothesis of rotation invariance falls down. Thus, the image-based and text-based bases are not equal Bell states as defined in Eq. (4).

This result may be related to our experimental setting that the outcomes of the observables are initially independent. For instance, the probability of the text-based relevance of the first document does not affect the probability of the text-based relevance of the second document. Thus, the joint probability of relevance is calculated as a product of individual relevance probabilities. However, in [1, 2, 6, 7] the Bell inequality has been violated. In those experiments, the users are asked to report their judgments on composite states. Hence the joint probabilities can be directly estimated from the judgments. Thus, the expectation values are calculated under an implicit assumption that the outcomes can be incompatible. This assumption may result in "conjunction fallacy" [20] violating the monotonicity law of probability by overestimating the joint probability, thus violating the Bell inequality.

Our result may be also due to the dataset that has been used to conduct the experiment. In ImageClef2007, the outcomes are independent, i.e., the text-based and image-based relevance, therefore we cannot make the opposite assumption. Thus, we may need another dataset containing relevance judgment for a pair of documents. Additionally, we may search for a dataset where Bell states (i.e., Eq. (2)) preexist, such that an interaction between two documents cannot be validly decomposed and modeled as interaction of separate documents. Then, the Bell inequality may be violated for those cases.

Finally, we experimentally investigated the violation of the Bell inequality in a small-scale experiment. In the current experiment, for each query, we focused on a small amount of relevant and irrelevant multimodal documents trying to search for non-classical correlations between two documents. However, it is worth conducting a large-scale experiment as well, looking also at a general first round retrieval process, or even at relevance feedback scenario. Moreover, it would be interesting to investigate the existence of non-classical correlations among many documents. Then, the CHSH inequality should be generalized for systems with multiple settings or basis [10].

7   Conclusion

In this paper, we have investigated non-classical correlations between pairs of decision fused multimodal documents. We examined the existence of such correlations through the violation of the CHSH inequality. In this case, a violation implies that measuring a mono-modal decision in a document, we could instantaneously predict with certainty a mono-modal decision in the other system acquiring information about how to fuse local decisions. Unfortunately, we did not find any violation of the Bell inequality. This result may be related to our assumption that the outcomes of the observables are initially independent. The result may also be due to the dataset. On one hand there is no real user involved in relevance judgment; on the other hand there do not exist initial Bell states between two multimodal documents. Nevertheless, the experimental results and discussions may provide theoretical and empirical insights and inspirations for future development of this direction.

Acknowledgement. This work is funded by the European Union's Horizon 2020 research and innovation programme under the Marie Sklodowska-Curie grant agreement No 721321.

A   Appendix

The expectation of a random variable X that takes the values $\{+, -\}$ according to the probability distribution $P_{X(+)}, P_{X(-)}$ is defined as

$$\langle X \rangle = (+)P_{X(+)} + (-)P_{X(-)}.$$

For two random variables X, Ψ, that take the values $\{+, -\}$ according to the probability distribution $P_{X(+)}, P_{X(-)}$ and $P_{\Psi(+)}, P_{\Psi(-)}$ respectively, the expectation value is defined as the product resulting in

$$\begin{aligned}
\langle X, \Psi \rangle &= ((+)P_{X(+)} + (-)P_{X(-)}) * ((+)P_{\Psi(+)} + (-)P_{\Psi(-)}) \\
&= (+)(+)(P_{X(+)}P_{\Psi(+)}) + (+)(-)(P_{X(+)}P_{\Psi(-)}) \\
&\quad + (-)(+)(P_{X(-)}P_{\Psi(+)}) + (-)(-)(P_{X(-)}P_{\Psi(-)}).
\end{aligned}$$

References

1. Aerts, D., Sozzo, S.: Quantum structure in cognition: why and how concepts are entangled. In: Song, D., Melucci, M., Frommholz, I., Zhang, P., Wang, L., Arafat, S. (eds.) QI 2011. LNCS, vol. 7052, pp. 116–127. Springer, Heidelberg (2011). https://doi.org/10.1007/978-3-642-24971-6_12
2. Aerts, D., Sozzo, S.: Quantum entanglement in concept combinations. Int. J. Theor. Phys. **53**(10), 3587–3603 (2014)
3. Aspect, A., Grangier, P., Roger, G.: Experimental realization of Einstein-Podolsky-Rosen-Bohm Gedankenexperiment: a new violation of Bell's inequalities. Phys. Rev. Lett. **49**(2), 91 (1982)
4. Atrey, P.K., Hossain, M.A., El Saddik, A., Kankanhalli, M.S.: Multimodal fusion for multimedia analysis: a survey. Multimedia Syst. **16**(6), 345–379 (2010)
5. Baltrušaitis, T., Ahuja, C., Morency, L.P.: Multimodal machine learning: a survey and taxonomy. IEEE Trans. Pattern Anal. Mach. Intell. **41**, 423–443 (2018)
6. Bruza, P.D., Kitto, K., Ramm, B., Sitbon, L., Song, D., Blomberg, S.: Quantum-like non-separability of concept combinations, emergent associates and abduction. Logic J. IGPL **20**(2), 445–457 (2011)

7. Bruza, P.D., Kitto, K., Ramm, B.J., Sitbon, L.: A probabilistic framework for analysing the compositionality of conceptual combinations. J. Math. Psychol. **67**, 26–38 (2015)
8. Cirel'son, B.S.: Quantum generalizations of Bell's inequality. Lett. Math. Phys. **4**(2), 93–100 (1980)
9. Clauser, J.F., Horne, M.A., Shimony, A., Holt, R.A.: Proposed experiment to test local hidden-variable theories. Phys. Rev. Lett. **23**(15), 880 (1969)
10. Gisin, N.: Bell inequality for arbitrary many settings of the analyzers. Phys. Lett. A **260**(1–2), 1–3 (1999)
11. Gleason, A.M.: Measures on the closed subspaces of a Hilbert space. J. Math. Mech. **6**, 885–893 (1957)
12. Grubinger, M., Clough, P., Hanbury, A., Müller, H.: Overview of the ImageCLEF-photo 2007 photographic retrieval task. In: Peters, C., et al. (eds.) CLEF 2007. LNCS, vol. 5152, pp. 433–444. Springer, Heidelberg (2008). https://doi.org/10.1007/978-3-540-85760-0_57
13. Hou, Y., Song, D.: Characterizing pure high-order entanglements in lexical semantic spaces via information geometry. In: Bruza, P., Sofge, D., Lawless, W., van Rijsbergen, K., Klusch, M. (eds.) QI 2009. LNCS (LNAI), vol. 5494, pp. 237–250. Springer, Heidelberg (2009). https://doi.org/10.1007/978-3-642-00834-4_20
14. Hou, Y., Zhao, X., Song, D., Li, W.: Mining pure high-order word associations via information geometry for information retrieval. ACM Trans. Inf. Syst. (TOIS) **31**(3), 12 (2013)
15. Melucci, M.: Introduction to Information Retrieval and Quantum Mechanics, pp. 156–158, 176–181, 212–213, 217–221. Springer, Berlin (2015). https://doi.org/10.1007/978-3-662-48313-8
16. Nielsen, M.A., Chuang, I.: Quantum computation and quantum information (2002)
17. Pathak, A.: Elements of Quantum Computation and Quantum Communication, pp. 92–98. Taylor & Francis, Abingdon (2013)
18. Simonyan, K., Zisserman, A.: Very deep convolutional networks for large-scale image recognition. arXiv preprint arXiv:1409.1556 (2014)
19. Stenger, V.J.: Timeless Reality: Symmetry, Simplicity and Multiple Universes. (Chap. 12)
20. Tversky, A., Kahneman, D.: Extensional versus intuitive reasoning: the conjunction fallacy in probability judgment. Psychol. Rev. **90**(4), 293 (1983)
21. Van Rijsbergen, C.J.: The Geometry of Information Retrieval. Cambridge University Press, Cambridge (2004)
22. Veloz, T., Zhao, X., Aerts, D.: Measuring conceptual entanglement in collections of documents. In: Atmanspacher, H., Haven, E., Kitto, K., Raine, D. (eds.) QI 2013. LNCS, vol. 8369, pp. 134–146. Springer, Heidelberg (2014). https://doi.org/10.1007/978-3-642-54943-4_12

Investigating Bell Inequalities
for Multidimensional Relevance
Judgments in Information Retrieval

Sagar Uprety[1][✉], Dimitris Gkoumas[1], and Dawei Song[1,2]

[1] The Open University, Milton Keynes, UK
{sagar.uprety,dimitris.gkoumas,dawei.song}@open.ac.uk
[2] Beijing Institute of Technology, Beijing, China

Abstract. Relevance judgment in Information Retrieval is influenced by multiple factors. These include not only the topicality of the documents but also other user oriented factors like trust, user interest, etc. Recent works have identified and classified these various factors into seven dimensions of relevance. In a previous work, these relevance dimensions were quantified and user's cognitive state with respect to a document was represented as a state vector in a Hilbert Space, with each relevance dimension representing a basis. It was observed that relevance dimensions are incompatible in some documents, when making a judgment. Incompatibility being a fundamental feature of Quantum Theory, this motivated us to test the Quantum nature of relevance judgments using Bell type inequalities. However, none of the Bell-type inequalities tested have shown any violation. We discuss our methodology to construct incompatible basis for documents from real world query log data, the experiments to test Bell inequalities on this dataset and possible reasons for the lack of violation.

Keywords: Quantum cognition · Information Retrieval · Multidimensional relevance · Bell inequalities

1 Introduction

Information Retrieval (IR) is defined as finding material (documents, videos, audio, etc.) of an unstructured nature that are relevant to an information need of the user. Information Need (IN) of a user is usually expressed as a query. An essential component of IR is the concept of relevance of documents. It is defined as how well a document satisfies the user Information Need. Relevance in IR was traditionally considered to be Topical, i.e. how well the content of the retrieved document matches the topic of the query(e.g. text match). As content similarity matching techniques have become more accurate, almost all of the documents obtained for a query generally satisfy the topicality criteria. Hence users tend to consider other factors while judging documents. These different factors have been investigated in several works [3,20,21]. In [13], seven relevance

© Springer Nature Switzerland AG 2019
B. Coecke and A. Lambert-Mogiliansky (Eds.): QI 2018, LNCS 11690, pp. 177–188, 2019.
https://doi.org/10.1007/978-3-030-35895-2_12

dimensions were identified. Each of these dimensions was quantified by defining certain features, which could be extracted from all query-document pairs. These seven dimensions are described in Table 1.

Table 1. Seven dimensions of relevance

Relevance dimensions	
Topicality	The extent to which the retrieved document is related to the topic of the current query
Reliability	The degree to which the content of the document is true, accurate and believable. Determined by the reliability of source
Understandability	Extent to which the contents are readable. Vocabulary, complexity of sentences, layout of pages, etc. taken into consideration
Interest	Topics from user's past searches
Habit	Focus on behavioral preference of users, e.g. always using certain websites for particular tasks
Scope	Whether both breadth and depth of the document are suitable to the Information Need
Novelty	Whether the document contains information which is new to the user, or the document itself is newly created

In [19], a document defined using these relevance dimensions is represented as a two dimensional Hilbert space. Each of the seven relevance dimensions is represented as a basis. The different basis correspond to the different perspectives of relevance judgment for the same document. Based on which relevance dimension is considered, the same document will have different probabilities of relevance. Thus the document exists in multiple states (e.g. highly relevant, not relevant, moderately relevant, etc.) simultaneously and we get a particular judgment depending upon which criteria (relevance dimension) the user used to judge (measure) it. This is analogous to the measurement of electron spin which is either up or down in direction, but depends upon which axis it is measured in. Electrons with spin up along the Z-axis may have both up and down components along the X-axis. So a document may look relevant based on the Topicality dimension, but may not be so along, say, the Reliability dimension. We discuss the methodology used to quantify these seven dimensions and construct Hilbert spaces for documents in the next section.

This incompatibility in judgment perspectives is a fundamental feature of Quantum Mechanics [15]. Incompatibility forbids the possibility of jointly determining the outcome of an event from two perspectives. We investigate whether decision making in IR, consisting of multiple perspectives, has an analogous quantum phenomena. A formal test of quantumness of systems was given in

1964 by Bell [5]. He formulated an inequality which cannot be violated by classical systems governed by joint probability distributions. Quantum Mechanics was shown to violate it for particular settings. In this work, we use another version of the Bell inequality, called the CHSH inequality [10]. The CHSH inequality is given by Eq. (1) for two systems A and B where observables A_1 and A_2 can be measured in system A and B_1 and B_2 can be measured in system B. A_i and B_i can take values only in $\{\pm 1\}$. It is assumed that the observables have pre-existing values which are not influenced by any other measurement.

$$|\langle A_1 B_1 \rangle + \langle A_1 B_2 \rangle + \langle A_2 B_1 \rangle - \langle A_2 B_2 \rangle| \leq 2 \tag{1}$$

The CHSH inequality is violated in Quantum Mechanics using a special composite state of two systems, called the Bell state [14], which has the following form:

$$|\psi\rangle = \frac{1}{\sqrt{2}}(|00\rangle + |11\rangle) \tag{2}$$

where $|0\rangle$ and $|1\rangle$ represent the standard basis for the two systems. Initially, both the systems are in a superposed state. The two outcomes, i.e., corresponding to the $|0\rangle$ and $|1\rangle$ vectors can be obtained with equal probabilities. However, on measuring one system, if one obtains the outcome corresponding to the basis vector $|0\rangle$, the state of the composite system collapses to $|00\rangle$. Now it is known for certain that the outcome of the second system also corresponds to $|0\rangle$. This is true even if the two systems are spatially separated - the measurement on one system reveals the state of the other, instantaneously.

Violation of Bell inequalities by such entangled states prove the impossibility of the existence of a joint probability distribution for the two systems. It rules out the concept of "Local Realism" of the classical world, which is the assumption made while deriving the Bell inequalities. 'Local' implies the fact that measurement of one system does not influence that of a spatially separated system. 'Realism' assumes that values of physical properties of systems have definite values and exist independent of observation [14].

There have been several works which have investigated violation of Bell inequalities in macroscopic and cognitive systems [1,2,6]. This work also investigates the Bell inequalities for violation by user's composite state for judgment of two documents. After describing the methodology used to quantify the seven relevance dimensions, we describe equivalent Bell inequalities for the user states for documents. Subsequently we give details of the experimental settings used to form the composite system of documents.

2   Quantifying Relevance Dimensions

We represent each document as a two-dimensional real valued Hilbert space. The two basis vectors correspond to relevance and non-relevance of a dimension. For the seven dimensions, we have seven different basis in the Hilbert space. The user's cognitive state for this document is a vector in the Hilbert space, a superposition of the basis vectors. Using the Dirac notation, we get the user state for a document d in different basis as:

$$|d\rangle = \alpha_{11}|R_{hab}\rangle + \beta_{11}|\widetilde{R_{hab}}\rangle$$
$$= \alpha_{12}|R_{int}\rangle + \beta_{12}|\widetilde{R_{int}}\rangle \tag{3}$$

and so on, in all seven basis. The coefficients $|\alpha_x|^2$ is the weight (i.e., probability of relevance) the user assigns to document d in terms of the dimension x, and $|\alpha_x|^2 + |\beta_x|^2 = 1$.

To calculate the coefficients of superposition in a basis, we use the same technique as [?]. The dataset is of query logs from the Bing search engine. Following the methodology in [13], we define a set of features for each of the seven relevance dimensions. For each query-document pair, the set of features for each dimension are extracted and integrated into the LambdaMART [8] Learning to Rank (LTR) algorithm to generate seven relevance scores (one for each dimension) for the query-document pair. Due to lack of space, we refer the readers to [13] for more details on the features defined for each dimension and also how they are used in the LTR algorithm. We thus get seven different ranked lists for a query, corresponding to each relevance dimension. Then the scores assigned to a document for each dimension are normalized using the min-max normalization technique, across all the documents for the query. The normalized score for each dimension forms the coefficient of superposition of the relevance vector for the respective dimension. For example, for a query q, let $d_1, d_2, ..., d_n$ be the ranking order corresponding to the "Reliability" dimension, based on the relevance scores of $\lambda_1, \lambda_2, ..., \lambda_n$ respectively. We construct the vector for document d_1 in the 'Reliability' basis as:

$$|d_1\rangle = \alpha_{11}|R_{rel}\rangle + \beta_{11}|\widetilde{R_{rel}}\rangle \tag{4}$$

where $\alpha_{11} = \sqrt{\frac{\lambda_1 - min(\lambda)}{max(\lambda) - min(\lambda)}}$, where $max(\lambda)$ is the maximum value among $\lambda_1, \lambda_2, ..., \lambda_n$. Square root is taken to enable calculation of probabilities according to the Born rule. We can thus represent this document in all the seven basis and therefore all the documents in their respective Hilbert spaces.

For documents where α_{11} and α_{12} are different, we get incompatible basis. Incompatibility in relevance dimensions for judging documents can be manifested in terms of Order Effects. Different order of considering relevance dimensions while judging a document will lead to different final judgments. As an example, consider a document with the following Hilbert space [18]:

$$|d\rangle = 0.9715|R\rangle + 0.2370|\widetilde{R}\rangle \tag{5}$$

$$= 0.3535|T\rangle + 0.9354|\widetilde{T}\rangle \tag{6}$$

We take the *Reliability* basis($|R\rangle, |\widetilde{R}\rangle$)as the standard basis. Representing *Topicality* basis in the standard *Reliability* basis, we get (Appendix A):

$$|T\rangle = 0.5651|R\rangle + 0.8250|\widetilde{R}\rangle \tag{7}$$

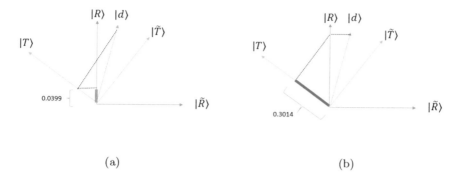

Fig. 1. Hilbert space representation of Order Effects

Suppose that while judging Document d, the user has the order $Topicality \rightarrow Reliability$ in mind. Then the final probability of relevance is the projection from $d \rightarrow T \rightarrow R$ as shown in Fig. 1a. This is calculated as $|\langle T| |d\rangle|^2 |\langle R| |T\rangle|^2 = 0.3535^2 * 0.5651^2 = 0.0399$. If the user reverses the order of relevance dimensions considered while judging document d, we get $d \rightarrow R \rightarrow T = |\langle R| |d\rangle|^2 |\langle T| |R\rangle|^2 = 0.9715^2 * 0.5651^2 = 0.3014$, which is 7.5 times larger (Fig. 1b).

Order Effects in decision making have been successfully modeled and predicted using the Quantum framework [7,16].

3   Deriving a Bell Inequality for Documents

3.1   CHSH Inequality

In Sect. 2, we showed how we can calculate the relevance probabilities of a document for different dimensions. We constructed a Hilbert space for each document, consisting of seven different basis, representing each dimension of relevance. Two or more such documents can be considered as a composite system by taking a tensor product of the document Hilbert spaces. If $|d_1\rangle$ and $|d_2\rangle$ are the state vectors of two documents, we can represent the tensor product as $|d_1\rangle \bigotimes |d_2\rangle$. Figure 2 shows the geometrical representation of two such Hilbert spaces. Here $|R\rangle_{hab}$ represents Relevance in the Habit basis, or in IR terms, relevance of document d with respect to the Habit dimension. Similarly, $|\tilde{R}\rangle_{hab}$ represents irrelevance in the Habit basis.

In the CHSH inequality, we have observables A_1 and A_2 for a system taking values in ± 1. For a document d_1, we have observables corresponding to the different relevance dimensions. Taking the case of two relevance dimensions, Habit and Novelty, we have observables R_{hab} and R_{nov} which take values in ± 1. Where $R_{hab} = +1$ corresponds to a projection on the basis vector $|R\rangle_{hab}$, $R_{hab} = -1$ corresponds to the projection on its orthogonal basis vector $|\tilde{R}\rangle_{hab}$.

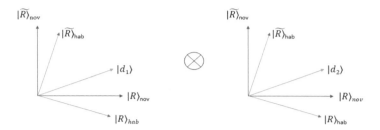

Fig. 2. Tensor product of two Hilbert spaces

Taking two documents as a composite system, we can write the CHSH inequality in the following way:

$$|\langle R_{hab1}R_{hab2}\rangle + \langle R_{hab1}R_{nov2}\rangle + \langle R_{nov1}R_{hab2}\rangle - \langle R_{nov1}R_{nov2}\rangle| \leq 2 \qquad (8)$$

where the subscripts 1 and 2 denote that the observables belong to document 1 and document 2 respectively. Using the fact that $\langle AB \rangle = 1 * P(AB = 1) + (-1) * P(AB = -1)$ and $P(AB = 1) + P(AB = -1) = 1$, we can convert the above inequality into its probability form as:

$$
\begin{aligned}
1 \quad \leq \quad & P(R_{hab1}R_{hab2} = 1) + P(R_{hab1}R_{nov2} = 1) + \\
& P(R_{nov1}R_{hab2} = 1) + P(R_{nov1}R_{nov2} = -1) \quad \leq \quad 3
\end{aligned}
\qquad (9)
$$

We don't have the joint probabilities $P(AB)$ in our dataset, hence we assuming $P(AB) = P(A)P(B)$ (this where the assumption of realism is incorrectly made, which will not lead to the CHSH inequality violation), we get:

$$
\begin{aligned}
1 \quad \leq \quad & P(R_{hab1} = 1)P(R_{hab2} = 1) + P(R_{hab1} = -1)P(R_{hab2} = -1) + \\
& P(R_{hab1} = 1)P(R_{nov2} = 1) + P(R_{hab1} = -1)P(R_{nov2} = -1) + \\
& P(R_{nov1} = 1)P(R_{hab2} = 1) + P(R_{nov1} = -1)P(R_{hab2} = -1) + \\
& P(R_{nov1} = 1)P(R_{nov2} = -1) + P(R_{nov1} = -1)P(R_{nov2} = 1) \quad \leq \quad 3
\end{aligned}
\qquad (10)
$$

As we mentioned above, $R_{hab} = +1$ corresponds to the basis vector $|R_{hab}\rangle$ and therefore $P(R_{hab1} = 1)$ corresponds to the probability that document d_1 is relevant with respect to the *Habit* dimension of relevance. Therefore we can calculate these probabilities as projections in the Hilbert space:

$$
\begin{aligned}
P(R_{hab1} = 1) \quad &= |\langle R_{hab}|d_1\rangle|^2 \\
P(R_{hab1} = -1) &= |\langle \widetilde{R_{hab}}|d_1\rangle|^2 \\
P(R_{nov1} = 1) \quad &= |\langle R_{nov}|d_1\rangle|^2 \\
P(R_{nov1} = -1) &= |\langle \widetilde{R_{nov}}|d_1\rangle|^2
\end{aligned}
\qquad (11)
$$

and similarly for document d_2.

3.2   CHSH Inequality for Documents Using the Trace Method

Another way to define the CHSH inequality for documents is by directly calculating the expectation values using the trace rule. According to this rule, expectation value of an observable A in a state $|d\rangle$ is given by

$$\langle A \rangle = tr(A\rho) \tag{12}$$

where the quantity $\rho = |d\rangle \langle d|$ is the density matrix for the state $|d\rangle$.

Let the two documents be represented in the standard basis as follows:

$$|D_1\rangle = a_1 |H\rangle_1 + b_1 |\tilde{H}\rangle_1$$
$$|D_2\rangle = a_2 |H\rangle_2 + b_2 |\tilde{H}\rangle_2 \tag{13}$$

where $|H\rangle_{1,2} = \begin{pmatrix} 1 \\ 0 \end{pmatrix}$ and $|\tilde{H}\rangle_{1,2} = \begin{pmatrix} 0 \\ 1 \end{pmatrix}$ Hence, the state vector and the density matrix for a document $|D\rangle$ can be written as:

$$|D\rangle = \begin{pmatrix} a \\ b \end{pmatrix} \qquad |D\rangle \langle D| = \begin{pmatrix} a_1^2 & a_1 b_1 \\ a_1 b_1 & b_1^2 \end{pmatrix} \tag{14}$$

The document representations in another basis are as follows:

$$|D_1\rangle = c_1 |N\rangle_1 + d_1 |\tilde{N}\rangle_1$$
$$|D_2\rangle = c_2 |N\rangle_2 + d_2 |\tilde{N}\rangle_2 \tag{15}$$

H and N are basically relevance with respect to two relevance dimensions, say Habit and Novelty. We can write the N basis in terms of the H basis (see Appendix A) as:

$$|N\rangle_1 = (a_1 c_1 + b_1 d_1) |H\rangle_1 + (b_1 c_1 - a_1 d_1) |\tilde{H}\rangle_1$$
$$|\tilde{N}\rangle_1 = (a_1 d_1 - b_1 c_1) |H\rangle_1 + (a_1 c_1 + b_1 d_1) |\tilde{H}\rangle_1 \tag{16}$$

and similarly for the second document.

Thus we get the vector representations for basis states $|N\rangle_1$ and $|\tilde{N}\rangle_1$ as:

$$|N\rangle_1 = \begin{pmatrix} a_1 c_1 + b_1 d_1 \\ b_1 c_1 - a_1 d_1 \end{pmatrix} \qquad |\tilde{N}\rangle_1 = \begin{pmatrix} a_1 d_1 - b_1 c_1 \\ a_1 c_1 + b_1 d_1 \end{pmatrix} \tag{17}$$

Now the observables $\boldsymbol{H}$ and $\boldsymbol{N}$ are defined as:

$$\boldsymbol{H} = |H\rangle \langle H| - |\tilde{H}\rangle \langle \tilde{H}| \tag{18}$$
$$\boldsymbol{N} = |H\rangle \langle N| - |\tilde{N}\rangle \langle \tilde{N}|$$

where $|H\rangle \langle H|$ and $|\tilde{H}\rangle \langle \tilde{H}|$ are the projection operators for standard basis vectors with eigen values 1 and -1 respectively. This is the spectral decomposition of the observables. We get $\boldsymbol{H} = \begin{pmatrix} 1 & 0 \\ 0 & -1 \end{pmatrix}$. The matrix for observable $\boldsymbol{N}$ is obtained in terms of the amplitudes a, b, c and d. Now the CHSH inequality for the observables $\boldsymbol{H}$ and $\boldsymbol{N}$ acting on the two documents can be written as:

$$|\langle \boldsymbol{H}_1\boldsymbol{H}_2\rangle + \langle \boldsymbol{H}_1\boldsymbol{N}_2\rangle + \langle \boldsymbol{N}_1\boldsymbol{H}_2\rangle - \langle \boldsymbol{N}_1\boldsymbol{N}_2\rangle| \leq 2 \tag{19}$$

Here $\boldsymbol{H}_1\boldsymbol{H}_2$ denotes that we measure the observable $\boldsymbol{H}$ on both the documents. In the language of tensor products,

$$\boldsymbol{H}_1 \otimes \boldsymbol{N}_2 |D_1\rangle \otimes |D_2\rangle = \boldsymbol{H}_1 |D_1\rangle \otimes \boldsymbol{N}_2 |D_2\rangle \tag{20}$$

And,

$$\begin{aligned}
\langle \boldsymbol{H}_1\boldsymbol{N}_2\rangle &= \langle D_1 \otimes D_2 | \boldsymbol{H}_1 \otimes \boldsymbol{N}_2 | D_1 \otimes D_2\rangle \\
&= \langle D_1 | \boldsymbol{H}_1 | D_1\rangle \langle D_2 | \boldsymbol{N}_2 | D_2\rangle \\
&= tr(\boldsymbol{H}_1 |D_1\rangle \langle D_1|) \times tr(\boldsymbol{N}_2 |D_2\rangle \langle D_2|)
\end{aligned} \tag{21}$$

In this way we can directly calculate the expectation values in Eq. (19). As a sample calculation, $tr(\boldsymbol{H}_1 |D_1\rangle \langle D_1|) = tr\left(\begin{pmatrix} 1 & 0 \\ 0 & -1 \end{pmatrix} \begin{pmatrix} a_1^2 & a_1 b_1 \\ a_1 b_1 & b_1^2 \end{pmatrix} \right) = a_1^2 - b_1^2$, where a_1^2 and b_1^2 are the probabilities of relevance and non-relevance respectively in the standard (Habit) basis.

3.3   N-Settings Bell Inequality

The CHSH inequality refers to two two-dimensional systems where each system has two measurement settings (or measurement basis). However this can be generalized for systems with multiple settings or basis [11]

$$\sum_{j=1}^{n} \left(\sum_{k=1}^{n+1-j} E(A_j B_k) - \sum_{k=n+2-j}^{n} E(A j B_k) \right) \leq \left[\frac{n^2 + 1}{2} \right] \tag{22}$$

where $[x]$ denotes the largest integer smaller or equal to x.

For seven relevance dimensions, $n = 7$ and the bound is 25. We can convert Eq. (22) into its probability form as done in Sect. 3.1, or use the trace rule to directly calculate the expectation values as done in Sect. 3.2

4   Experiment and Results

Having obtained an equivalent representation of Bell inequalities in Sect. 3, we proceed to substitute the values in the inequalities and test for violation using relevance scores as calculated in Sect. 2. For each query, a user judges several documents to be relevant or non-relevant according to his or her information need. We investigate the correlations between these documents, with each document having multiple decision perspectives, using the Bell Inequalities. We consider the following types of document pairs to test for Quantum Correlations:

(I) We consider those queries where only two documents are SAT clicked (Satisfied Click - Those documents which are clicked and browsed for at least 30 s). Out of 55617 queries in our dataset, 1702 queries had exactly two SAT clicked documents. We consider a composite system of these two documents and measure (judge the relevance) along different basis (relevance dimensions) corresponding to each of the Bell inequalities described in Sects. 3.1, 3.2 and 3.3.

(II) We consider those queries for which we have at least one SAT clicked document. Out of 55617 queries in our dataset, we find 52936 queries with at least one SAT clicked document. We then consider a composite system of this SAT clicked document with all the unclicked documents for the query (one by one) and measure (judge the relevance) along different basis (relevance dimensions) corresponding to each of the Bell inequalities described in Sects. 3.1, 3.2, 3.3. In both cases, we do not find the violation of the Bell inequalities for any query. While case (I) corresponds to correlated documents and case (II) corresponds to anti-correlated documents, it is to be noted that we are taking a composite system by taking a tensor product of two document states. This, in turn is separable back into the two document states. The reason why Quantum Mechanics violates Bell Inequalities is due to the existence of non-separable states like the Bell States. To get something similar to an entangled state, we consider another type of document pairs:

(III) Consider a pair of documents which are listed together for many queries, but are always judged in a correlated manner. That is, if one document of the pair is SAT clicked, the other one is also SAT clicked for that query. And similarly both might be unclicked for another query in which they appear together. Also, we find those documents which are SAT clicked together in half of the queries they occur in, and unclicked in the other half. This corresponds to the following Bell State:

$$|\psi\rangle = \frac{1}{\sqrt{2}}(|RR\rangle + |\widetilde{R}\widetilde{R}\rangle) \tag{23}$$

We take such pairs of documents to test the Bell inequalities on them. Out of 774 pairs of documents, no pair show the violation of the inequalities discussed above.

The composite state of the two documents described in Eq. (23) appears to be like an entangled state of the documents - knowing that one document is SAT clicked or not can tell us about the other document. However, one fundamental property of the Bell states is their rotational invariance. Representing a Bell State in any basis, one gets the same probabilities of the two possible outcomes. For example,

$$|\psi\rangle = \frac{1}{\sqrt{2}}\left(|HH\rangle + |\widetilde{H}\widetilde{H}\rangle\right) \tag{24}$$

$$= \frac{1}{\sqrt{2}}\left(|TT\rangle + |\widetilde{T}\widetilde{T}\rangle\right)$$

where H, N and T are relevance with respect to the Habit, Topicality and Novelty basis. One can always hypothetically construct document Hilbert spaces in such a manner that the composite state is rotationally invariant, but that is not the case in the query log data, which is the target of our investigation.

As a formal test of non-separable states, we perform Schmidt decomposition [14] of the composite system of document pairs. We do not find any evidence of non-separable states for any type of document pairs, as described in cases (I), (II) and (III).

5   Conclusion and Future Work

We tested Bell inequalities for violation using data from Bing Query logs. Despite the presence of incompatible measurements, Bell inequalities are not violated. However, the incompatibility in measurement applies to the user's cognitive state with respect to a single document. Hence there might exist a joint probability distribution governing user's cognitive state for a pair of documents. The experiments in which the violation of Bell inequality has been reported for cognitive systems, the users are asked to report their judgments on composite states. Hence the joint probabilities can be directly estimated from the judgments. This may result in a "Conjunction Fallacy" [17] due to incompatible decision perspectives, thus violating the monotonicity law of probability by overestimating the joint probability, and therefore violating the Bell inequality. In our dataset, we don't have judgments over the document pairs. That is, the user does not judge a pair of document to be relevant with respect to some dimensions. Instead we have got the probabilities of relevance of a single document with respect to different dimensions. When we use the relevance probability of individual documents to compute the joint probabilities for a pair of documents, we are forced to assume the existence of a joint probability distribution. Thus there might be a possibility of Bell inequality violation if we can obtain data for a pair of documents. For example, users can be asked to rate a document to be relevant with respect to Novelty and another document relevant with respect to Topicality. This would correspond to the $E(R_{nov}, R_{top})$ term in the CHSH inequality. In this case, user's judgment of a document may affect judgment of the other document in the pair.

Another test of the quantum nature of relevance judgments can be to test the non-contextual inequalities like the KCBS inequality [12]. Bell inequalities are designed for a composite system with the assumption of locality and realism. The non-contextual inequalities are designed for a single system with multiple measurement perspectives, some of which are incompatible with each other. However, contextuality only exhibits in systems of more than two dimensions. Hence we need to modify our two-dimensional (two decision outcomes - relevant or not relevant) approach to test inequalities like the KCBS inequality. One can also test for violation of the Contextuality-by-Default inequality [4,9]. This forms part of our future work.

Acknowledgment. This work is funded by the European Union's Horizon 2020 research and innovation programme under the Marie Sklodowska-Curie grant agreement No 721321. We would like to thank Jingfei Li for his help in providing the processed dataset.

A   Appendix

Consider a state vector in two different basis of a two dimensional Hilbert space, $|\psi\rangle = a|A\rangle + b|B\rangle = c|C\rangle + d|D\rangle$ We want to represent the vectors of one basis in terms of the other. To do that, consider the vector orthogonal to $|\psi\rangle$, which is $|\widetilde{\psi}\rangle = b|A\rangle - a|B\rangle = d|C\rangle - c|D\rangle$

Using the above representations, we get

$$|C\rangle = c|\psi\rangle + d|\widetilde{\psi}\rangle \, and \, |D\rangle = d|\psi\rangle - c|\widetilde{\psi}\rangle \tag{25}$$

Substituting $|\psi\rangle = a|A\rangle + b|B\rangle$ and $|\widetilde{\psi}\rangle = b|A\rangle - a|B\rangle$ in 25, we get:

$$|C\rangle = (ac + bd)|A\rangle + (bc - ad)|B\rangle$$
$$|D\rangle = (ad - bc)|A\rangle + (ac + bd)|B\rangle \tag{26}$$

References

1. Aerts, D.: Found. Phys. **30**(9), 1387–1414 (2000). https://doi.org/10.1023/a:1026449716544
2. Aerts, D., Sozzo, S.: Quantum entanglement in concept combinations. Int. J. Theoret. Phys., pp. 3587–3603. (2013). https://doi.org/10.1007/s10773-013-1946-z
3. Barry, C.L.: Document representations and clues to document relevance. J. Am. Soc. Inf. Sci. **49**(14), 1293–1303 (1998)
4. Basieva, I., Cervantes, V.H., Dzhafarov, E.N., Khrennikov, A.: True contextuality beats direct influences in human decision making (2018)
5. Bell, J.S.: On the Einstein Podolsky Rosen paradox. Phys. Phys. Fiz. **1**, 195–200 (1964). https://doi.org/10.1103/PhysicsPhysiqueFizika.1.195
6. Bruza, P.D., Kitto, K., Ramm, B., Sitbon, L., Song, D., Blomberg, S.: Quantum-like non-separability of concept combinations, emergent associates and abduction. Logic J. IGPL **20**(2), 445–457 (2011). https://doi.org/10.1093/jigpal/jzq049
7. Bruza, P., Chang, V.: Perceptions of document relevance. Front. Psychol. **5**, 612 (2014). https://doi.org/10.3389/fpsyg.2014.00612
8. Burges, C.J.C.: From ranknet to lambdarank to lambdamart: an overview (2010)
9. Cervantes, V.H., Dzhafarov, E.N.: Snow queen is evil and beautiful: experimental evidence for probabilistic contextuality in human choices. Decision **5**(3), 193–204 (2018). https://doi.org/10.1037/dec0000095
10. Clauser, J.F., Horne, M.A., Shimony, A., Holt, R.A.: Proposed experiment to test local hidden-variable theories. Phys. Rev. Lett. **23**, 880–884 (1969). https://doi.org/10.1103/PhysRevLett.23.880
11. Gisin, N.: Bell inequality for arbitrary many settings of the analyzers. Phys. Lett. A **260**(1–2), 1–3 (1999). https://doi.org/10.1016/s0375-9601(99)00428-4
12. Klyachko, A.A., Can, M.A., Binicioğlu, S., Shumovsky, A.S.: Simple test for hidden variables in spin-1 systems. Phys. Rev. Lett. **101**, 020403 (2008). https://doi.org/10.1103/PhysRevLett.101.020403
13. Li, J., Zhang, P., Song, D., Wu, Y.: Understanding an enriched multidimensional user relevance model by analyzing query logs. J. Assoc. Inf. Sci. Technol. **68**(12), 2743–2754 (2017). https://doi.org/10.1002/asi.23868

14. Nielsen, M.A., Chuang, I.L.: Quantum Computation and Quantum Information: 10th Anniversary Edition, 10th edn. Cambridge University Press, New York (2011)
15. Sakurai, J.J., Napolitano, J.: Modern Quantum Mechanics. Cambridge University Press, Cambridge (2017)
16. Trueblood, J.S., Busemeyer, J.R.: A quantum probability account of order effects in inference. Cogn. Sci. **35**(8), 1518–1552 (2011). https://doi.org/10.1111/j.1551-6709.2011.01197.x
17. Tversky, A., Kahneman, D.: Extensional versus intuitive reasoning: the conjunction fallacy in probability judgment. Psychol. Rev. **90**(4), 293–315 (1983). https://doi.org/10.1037/0033-295x.90.4.293
18. Uprety, S., Song, D.: Investigating order effects in multidimensional relevance judgment using query logs. In: Proceedings of the 2018 ACM SIGIR International Conference on Theory of Information Retrieval, ICTIR 2018, pp. 191–194. ACM, New York (2018). https://doi.org/10.1145/3234944.3234972
19. Uprety, S., Su, Y., Song, D., Li, J.: Modeling multidimensional user relevance in IR using vector spaces. In: The 41st International ACM SIGIR Conference on Research and Development in Information Retrieval, SIGIR 2018, pp. 993–996. ACM, New York (2018). https://doi.org/10.1145/3209978.3210130
20. Xu, Y.C., Chen, Z.: Relevance judgment: what do information users consider beyond topicality? J. Am. Soc. Inf. Sci. Technol. **57**(7), 961–973 (2006). https://doi.org/10.1002/asi.20361
21. Zhang, Y., Zhang, J., Lease, M., Gwizdka, J.: Multidimensional relevance modeling via psychometrics and crowdsourcing. In: Proceedings of the 37th International ACM SIGIR Conference on Research & Development in Information Retrieval, SIGIR 2014, pp. 435–444. ACM, New York (2014). https://doi.org/10.1145/2600428.2609577

Short Paper

An Update on Updating

Bart Jacobs$^{(\boxtimes)}$

Institute for Computing and Information Sciences,
Radboud University, Nijmegen, The Netherlands
bart@cs.ru.nl

The main aim of this short contribution is to give an introduction to some challenging research issues wrt. updating and probabilistic logic, together with some relevant references. We use the word 'update' for what is also called 'belief update' or (probabilistic) 'conditioning'. It involves the adaptation of a probability distribution in the light of certain evidence. Such updating is typically expressed via conditional probabilities and is governed by Bayes' rule. It is a fascinating topic, with wide applications, ranging from statistical data analysis to cognition theory (see *e.g.* [3, 9]).

Updating exists both in classical probability and in quantum probability. One of the key characteristics of updating in a quantum setting is that it is *not* commutative: successive updates do not commute. This forms a basis for using quantum theory in cognition theory [1] since the human mind is also very sensitive to the order in which information is presented—or, in different words, to the order of priming. The study of quantum updating is still in its infancy, but already two different mathematical definitions have appeared, called 'lower' and 'upper' conditioning in [5], see also [2, 7]. Interestingly, the lower version satisfies the product rule, whereas the upper version satisfies Bayes' rule proper. Classically, there is no difference between these two rules (see [5] for details).

In classical probability things seem to be better understood. But that is only because in practice people mostly restrict themselves to *sharp* evidence, given by subsets of the space at hand. These subsets are used as predicates in updating. The situation changes when *soft* or *fuzzy* evidence is allowed, of the form: I was 80% sure that I heard the alarm. Updating with fuzzy evidence can be done basically in two ways, called 'constructive' (following Pearl) or 'destructive' (following Jeffrey), see [4]. Constructive and destructive updating agree on point evidence, but they can give completely different outcomes when applied with the same (soft) evidence (and the same prior). It is unclear which version of updating should be applied when. This is a bit worrying. Should we start asking our doctors: did you arrive at this most likely diagnosis via constructive or destructive updating?

Constructive updating involves a smooth integration of the prior distribution with the evidence, following the standard formula: *posterior* $\propto$ *prior · likelihood*. Constructive updating is commutative. It is such that if the evidence contains no information (is constant/uniform), then you learn nothing new from updating.

Destructive updating involves overriding the prior by the evidence. As a result, it is *not* commutative. If the evidence is what we can predict, then we learn nothing new from destructive updating. This also makes sense. Given the

© Springer Nature Switzerland AG 2019
B. Coecke and A. Lambert-Mogiliansky (Eds.): QI 2018, LNCS 11690, pp. 191–192, 2019.
https://doi.org/10.1007/978-3-030-35895-2

precise mathematical distinction between constructive and destructive updating in [4], the question also arises: which form of updating best matches cognitive experiments?

Thus, in the end we have four forms of updating: two quantum ones (lower and upper) and two classical ones (constructive and destructive). Clearly more research is needed to understand this situation. Part of this research should involve developing a proper probabilistic language for expressing logical and computational properties (see also [6]). It is an embarrasment to the field that no widely accepted and used probabilistic symbolic logic exists so far. Developing such a logic is by no means an easy task, for instance because probabilistic updating leads to non-monotonicity: adding assumptions may weaken the validity of the conclusion. Non-monotonicity is avoided by most logicians. However, it is quite natural in a probabilistic setting, as becomes clear in the quote below from [8] with which we conclude.

To those trained in traditional logics, symbolic reasoning is the standard, and nonmonotonicity a novelty. To students of probability, on the other hand, it is symbolic reasoning that is novel, not nonmonotonicity. Dealing with new facts that cause probabilities to change abruptly from very high values to very low values is a commonplace phenomenon in almost every probabilistic exercise and, naturally, has attracted special attention among probabilists. The new challenge for probabilists is to find ways of abstracting out the numerical character of high and low probabilities, and cast them in linguistic terms that reflect the natural process of accepting and retracting beliefs.

References

1. Busemeyer, J., Bruza, P.: Quantum Models of Cognition and Decision. Cambridge University Press, Cambridge (2012)
2. Coecke, B., Spekkens, R.: Picturing classical and quantum Bayesian inference. Synthese **186**(3), 651–696 (2012)
3. Hohwy, J.: The Predictive Mind. Oxford University Press, Oxford (2013)
4. Jacobs, B.: A mathematical account of soft evidence, and of Jeffrey's 'destructive' versus Pearl's 'constructive' updating. See arxiv.org/abs/1807.05609 2018
5. Jacobs, B.: Lower and upper conditioning in quantum Bayesian theory. In: Quantum Physics and Logic, EPTCS (2018)
6. Jacobs, B., Zanasi, F.: The logical essentials of Bayesian reasoning. See arxiv.org/abs/1804.01193, book chapter, to appear
7. Leifer, M., Spekkens, R.: Towards a formulation of quantum theory as a causally neutral theory of Bayesian inference. Phys. Rev. A **88**(5), 052130 (2013)
8. Pearl, J.: Probabilistic semantics for nonmonotonic reasoning: a survey. In: Brachman, R., Levesque, H., Reiter, R. (eds.) First International Conference on Principles of Knowledge Representation and Reasoning, pp. 505–516. Morgan Kaufmann, San Mateo (1989)
9. Sloman, S.: Causal Models. How People Think About the World and Its Alternatives. Oxford University Press, Oxford (2005)

Author Index

Printed in the United States
By Bookmasters